Groundwater and Seepage

M. E. Harr

School of Civil Engineering,
Purdue University

DOVER PUBLICATIONS, INC.
NEW YORK

Published in Canada by General Publishing Company, Ltd., 30 Lesmill
Road, Don Mills, Toronto, Ontario.
Published in the United Kingdom by Constable and Company, Ltd.,
3 The Lanchesters, 162–164 Fulham Palace Road, London W6 9ER.

This Dover edition, first published in 1991, is an unabridged and cor-
rected republication of the work first published by the McGraw-Hill Book
Company, New York, in 1962.

Manufactured in the United States of America
Dover Publications, Inc., 31 East 2nd Street, Mineola, N.Y. 11501

Library of Congress Cataloging-in-Publication Data

Harr, Milton Edward, 1925–
    Groundwater and seepage / Milton E. Harr.
        p.   cm.
    Originally published: New York : McGraw-Hill, 1962.
    Includes bibliographical references and index.
    ISBN 0-486-66881-9
    1. Groundwater flow.   2. Seepage.   I. Title.
GB1197.7.H36   1991
621.4—dc20   91-37779
                                                        CIP

To my wife

Preface

The movement of groundwater is a basic part of soil mechanics. Its influence can be found in almost every area of civil engineering, including irrigation and reclamation. In addition, the elegance and logical structure of its theory renders it of interest to engineering scientists and applied mathematicians.

The first objective of this book is to provide the engineer with an organized analytical approach to the solution of seepage problems. Physical flow systems with which the engineer must deal tend increasingly to require the use of *advanced engineering mathematics*—in particular, the theory of complex variables, conformal mapping, and elliptic functions. For this reason, appendixes have been included which endeavor to emphasize the usefulness of these concepts rather than their mathematical niceties. An effort has been made in the arrangement of the text to introduce these mathematical concepts gradually and with sufficient application to form a firm foundation for the framework of seepage theory. Thus, for example, in Chap. 4 and the first part of Chap. 5 the Schwarz-Christoffel transformation is amply developed and practiced before introducing elliptic functions.

The second objective is to make available in the English language the relatively little-known works of Russian theoreticians—in particular, N. N. Pavlovsky, P. Ya. Polubarinova-Kochina, F. B. Nelson-Skornyakov, V. V. Vedernikov, S. N. Numerov, and A. M. Mkhitatrian. To achieve this end, the author has made much use of the works of these and other Russian scholars. Suitable acknowledgments to original sources are provided throughout the text and are keyed to the references at the end of the book.

The form of this volume was designed to meet the needs of the undergraduate and graduate student and the practicing engineer. Most of the material has been given by the author in undergraduate elective and graduate courses in soil mechanics at Purdue University. Students enrolled in the courses have had backgrounds in civil engineering and are familiar with the fundamentals of soil mechanics. A knowledge of these general areas is assumed on the part of the reader of this volume. Students have found the study of the mathematical concepts to be particularly rewarding: however, in keeping with an utilitarian philosophy, a conscientious attempt has been made to reduce solutions to simple graphs or charts.

A number of completely worked examples are provided, and over 200 problems of varying degrees of difficulty are included.   These range from proofs of a routine nature to practical applications of the text material.

In writing a text of this type there arises the inevitable problem of selection.   The author's first impulse was toward the general theory of groundwater and seepage, including both steady and transient states. However, with the growth of the present volume, it was decided to consider only steady-state flow and to defer the transient problem to another volume.

The author wishes to express his gratitude to his friend and colleague Dr. Gerald A. Leonards for his invaluable suggestions and discussions, and to Mrs. J. Becknell for typing the manuscript.

M. E. Harr

Contents

List of Symbols

a = length
A = area; angle
A_p = area of pores

b = distance
B = angle
$B(\ )$ = beta function

c = constant
C = constant; angle
C_x = correction factor

d = diameter; length
D = diameter; length

e = void ratio; exponential (2.718 . . .)
$\mathbf{E}$ = complete elliptic integral of second kind
$E(m,\phi)$ = elliptic integral of second kind

$F(m,\phi)$ = elliptic integral of first kind
F.S. = factor of safety

g = gravitational constant (32.2 ft/sec)
G = Zhukovsky function

h = total head
h_c = height of capillary rise
H = horizontal contact (creep path); depth of water in ditch

$i = \sqrt{-1}$; hydraulic gradient
I = hydraulic gradient
I_{cr} = critical gradient
I_E = exit gradient

J = specific integrals

k = coefficient of permeability
k_0 = permeability
K = complete elliptic integral of first kind

L = length (generally horizontal)
L_{dr} = minimum length of drain
L_0 = distance to underdrain

m = modulus of elliptic integrals; slope; ratio of pore area
M = constant; moment

n = porosity; parameter; normal direction
N = constant of integration
N_e = number of equipotential drops
N_f = number of flow channels

p = pressure, parametric plane
P = pressure force

q = quantity of seepage (per unit normal to direction of flow)
$\bar{q}$ = reduced quantity of seepage
Q = total volume flow

r = radial distance; real axis of t plane
r_w = radius of well
R = Reynolds number; radius of influence
R_c = weighted creep ratio

s = distance; imaginary axis of t plane; depth of embedment of sheetpile
S = length of free surface
S_s = specific gravity

t = time; parametric plane; tangential direction
T = layer thickness
T_s = surface tension

u, v, w = components of discharge velocity in x, y, z directions
$\bar{u}, \bar{v}, \bar{w}$ = components of seepage velocity in x, y, z directions
U = velocity

V = volume; tangential velocity
V_v = volume of voids

w = complex potential
W = complex velocity

$y_0 = y$ intercept of basic parabola

$Z(\ ) = $ zeta function

$\alpha = $ angle; parameter
$\beta = $ angle; parameter
$\gamma = $ unit weight; angle
$\gamma_w = $ unit weight of water (62.4 pcf)
$\Gamma(\ ) = $ gamma function
$\Delta = $ horizontal projection of wetted slope
$\varepsilon = $ rate of infiltration or evaporation; small distance; parameter
$\zeta = $ distance
$\eta = $ head ratio
$\theta = $ Zhukovsky function; angle
$\lambda = $ distance
$\mu = $ coefficient of viscosity; parameter
$\nu = $ kinematic viscosity
$\Pi(m,n,x) = $ elliptic integral of third kind
$\Pi_0 = $ complete elliptic integral of third kind
$\rho = $ density; radial distance
$\sigma = $ distance
$\tau = $ parameter
$\phi = $ potential function (velocity potential); amplitude of elliptic integral
$\Phi = $ form factor
$\psi = $ stream function
$\Omega = $ gravity potential

1

Fundamentals of Groundwater Flow

1-1. Scope and Aim of Subject

The aim of this work is primarily to present to the civil engineer the means of predicting the exigencies arising from the flow of groundwater. The specific problems which are to be dealt with can be divided into three parts:

1. Estimation of the quantity of seepage
2. Definition of the flow domain
3. Stability analysis

When writing on a subject as broad as groundwater and seepage, it is necessary to presume a minimum level of attainment on the part of the reader. Hence it will be assumed that the reader has a working knowledge of both the calculus and the rudiments of soil mechanics. For example, the problem of the stability of an earthen slope subject to seepage forces will be considered as solved once the upper flow line has been located and the pore pressures can be determined at all points within the flow domain. The actual mechanics of estimating the factor of safety of the slope will be left to the reader. Several texts on the subject of soil mechanics can be found in the references [142, 145].* However, such factors as the determination of the uplift pressures under structures, exit gradients, and all pertinent seepage quantities will be considered to be within the scope of this book.

Although the fundamentals of groundwater flow were established more than a century ago, it is only within recent years that the subject has met with scientific treatment. As a result of the trial-and-error history of groundwater-flow theory, its literature is replete with empirical relationships for which *exact* solutions can be and have been obtained. Advocates of the empirical approach have long reasoned that the heterogeneous nature of soils is such that rigorous analyses are not practical. This is

---

* Figures in brackets refer to the references at the end of the book.

1

not so; as will be seen, much of the subject lends itself readily to theoretical analysis.

Recent developments in the science of soil mechanics coupled with more precise methods of subsurface soil explorations have provided engineers with greater insight into the behavior of earth structures subject to groundwater flow. The mathematics of the theory of functions of complex variables, once the arid theory of imaginary numbers, now allows the engineer to solve problems of otherwise overpowering complexity.

The engineer can now formulate working solutions which not only reflect the interaction of the various flow factors but also allow him to obtain a measure of the uncertainties of his design.

1-2. Nature of Soil Body

In groundwater problems the soil body is considered to be a continuous medium of many interconnected openings which serve as the fluid carrier. The nature of the pore system within the soil can best be visualized by inference from the impermeable boundaries composing the *pore skeleton*. For simplicity it will be assumed that all soils can be divided into two fractions which will be referred to respectively as sand and clay.

In general, sands are composed of macroscopic particles that are *rounded* (bulky) or *angular* in shape. They drain readily, do not swell, possess insignificant capillary potential (see Sec. 1-7), and when dry exhibit no shrinkage. Clays, on the other hand, are composed of microscopic particles of platelike shape. They are highly impervious, exhibit considerable swelling, possess a high capillary potential, and demonstrate considerable volume reductions upon drying.

Sands approach more nearly the ideal porous medium and are representative of the soils primarily to be dealt with in this book. This may appear to introduce serious restrictions; however, in most engineering problems the low permeability of the clays renders them relatively impervious in comparison to the coarser-grained soils.

Extensive studies have been undertaken by many investigators [13, 45, 75, 84, 136] to calculate the permeability and porosity of natural soils based on their sieve analyses and various packings of uniform spheres. While it is not possible to derive significant permeability estimates from porosity measurements alone [86–88, 99], the pore characteristics of these ideal packings do present some of the salient features of natural soils.

Let us assume that the soil particles are all of uniform spherical shape. Calling the total volume V and the volume of voids V_v, we have for the *porosity*

$$n = \frac{V_v}{V} \tag{1}$$

and for the *void ratio*

$$e = \frac{V_v}{V - V_v} \qquad (2)$$

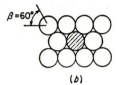

For a cubical array of spheres (Fig. 1-1a), $V = d^3$, $V_v = d^3 - \pi d^3/6$, and

$$n = 1 - \frac{\pi}{6} = 0.476$$

Fig. 1-1

For a rhombohedral packing (Fig. 1-1b), which represents the most compact assemblage of uniform spheres, the porosity is

$$n = 1 - \frac{\sqrt{2}}{6} \pi = 0.26$$

Values for the porosity of some natural soils are given in Table 1-1.

Table 1-1. Porosity of Some Natural Soils*

Description	Porosity
Uniform sand, loose	46
Uniform sand, dense	34
Glacial till, very mixed-grained	20
Soft glacial clay	55
Stiff glacial clay	37
Soft very organic clay	75
Soft bentonite	84

* From C. Terzaghi and R. B. Peck, "Soil Mechanics in Engineering Practice," p. 29, John Wiley & Sons, Inc., New York, 1948.

Figures 1-2a and b show the pore volume available for flow for the cubic and rhombohedral array, respectively [45]. It should be noted

from these figures that even in the ideal porous medium the pore space is not regular but consists of cavernous cells interconnected by narrower channels. Natural soils contain particles that can deviate considerably from the idealized spherical shape (as in the case of clay) and, in addition, are far from uniform in size. The true nature of the pore channels in a soil

Fig. 1-2. (*After Graton and Fraser* [45].)

mass defies rational description. It is for this reason that groundwater flow was not amenable to a scientific treatment until the advent of Darcy's law (Sec. 1-4). Fortunately, in groundwater problems we need not concern ourselves with the flow through individual channels. We are primarily interested in macroscopic flow wherein the flow across a section of many pore channels may be considered uniform as contrasted to the near-parabolic distribution of the flow through a single pore.

1-3. Discharge Velocity and Seepage Velocity

The *discharge velocity* is defined as the quantity of fluid that percolates through a unit of total area of the porous medium in a unit time.

As flow can occur only through the interconnected pores of saturated soils (Fig. 1-3), the velocity across any section must be thought of in a statistical sense. If m is the effective ratio of the area of pores A_p to the total area A, then $m = A_p/A$. The quantity of flow (also called the *quantity of seepage, discharge quantity,* or *discharge*) becomes $Q = mA\bar{v}$, where $m\bar{v}$ is the discharge velocity and $\bar{v}$ is called the *seepage velocity.*

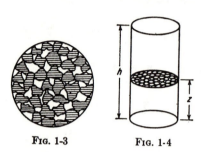

Fig. 1-3   Fig. 1-4

Let us investigate the nature of m. Designating $A_p(z)$ as the area of pores at any elevation z (Fig. 1-4), we have

$$m(z) = \frac{A_p(z)}{A} \tag{1}$$

The average value of m over the cylinder of height h is

$$m = \frac{1}{h} \int_0^h m(z) \, dz \tag{2}$$

Then

$$m = \frac{1}{Ah} \int_0^h A m(z) \, dz = \frac{1}{V} \int_0^h A_p(z) \, dz \tag{3}$$

where V is the total volume and the integral is the volume of voids. Hence the average value of m is the volume porosity n and will be so designated. In our work we shall deal mainly with discharge velocities, i.e., superficial velocities, and unless otherwise stated all velocities will be so inferred.

1-4. Darcy's Law

As is well known from fluid mechanics, for steady* flow of nonviscous incompressible fluids, Bernoulli's equation† [80]

$$\frac{p}{\gamma_w} + z + \frac{\bar{v}^2}{2g} = \text{constant} = h$$

where p = pressure, psf

γ_w = unit weight of fluid, pcf

---

* "Steady" denotes no time variation in fluid properties at a given point.

† The terms of Bernoulli's equation actually have the dimensions of energy per weight of fluid, foot-pounds per pound weight, or, as used in civil engineering, feet.

$\bar{v}$ = seepage velocity, ft/sec
g = gravitational constant, 32.2 ft/sec²
h = total head, ft

demonstrates that the sum of the *pressure head* p/γ_w, *elevation head* z, and *velocity head* $\bar{v}^2/2g$ at any point within the region of flow is a constant. In groundwater flow, to account for the loss of energy due to the viscous

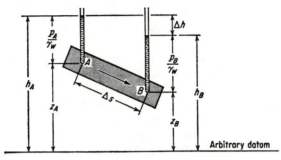

FIG. 1-5

resistance within the individual pores, Bernoulli's equation is taken as (Fig. 1-5)

$$\frac{p_A}{\gamma_w} + z_A + \frac{\bar{v}_A{}^2}{2g} = \frac{p_B}{\gamma_w} + z_B + \frac{\bar{v}_B{}^2}{2g} + \Delta h \qquad (1)$$

where Δh represents the total head loss (energy per unit weight of fluid) of the fluid over the distance Δs. The ratio

$$i = -\lim_{\Delta s \to 0} \frac{\Delta h}{\Delta s} = -\frac{dh}{ds} \qquad (2)$$

is called the *hydraulic gradient* and represents the space rate of energy dissipation per unit weight of fluid (a pure number).

In most groundwater problems the velocity heads (kinetic energy) are so small that they can be neglected. For example, a velocity of 1 ft/sec, which is large as compared to typical seepage velocities, produces a velocity head of only 0.015 ft. Hence, Eq. (1) can be written

$$\frac{p_A}{\gamma_w} + z_A = \frac{p_B}{\gamma_w} + z_B + \Delta h$$

and the total head at any point in the flow domain is simply

$$h = \frac{p}{\gamma_w} + z \qquad (3)$$

Prior to 1856, the formidable nature of groundwater flow defied rational analysis. In that year, Henry Darcy [25] published a simple relation based on his experiments on "les fontaines publiques de la ville de Dijon," namely,

$$v = ki = -k\frac{dh}{ds} \tag{4}$$

Equation (4), commonly called *Darcy's law*, demonstrates a linear dependency between the hydraulic gradient and the discharge velocity v. In soil mechanics, the coefficient of proportionality k is called the *coefficient of permeability* and, as shown in Eq. (4), k has the dimensions of a velocity.

Although Darcy's law is presented in differential form in Eq. (4), it must be emphasized that in no way does it describe the state of affairs within an individual pore. Strictly speaking, Darcy's law represents the *statistical macroscopic equivalent* of the Navier-Stokes equations of motion [Eqs. (1), Sec. 6-4] for the viscous flow of groundwater. It is precisely this equivalency that permits the subsequent development of groundwater flow within the theoretical framework of potential flow (Sec. 1-10). Stated simply, viscous effects are accounted for completely by Darcy's law, and hence for all subsequent considerations the flow may be treated as *nonviscous* or *frictionless*.

1-5. Range of Validity of Darcy's Law

Visual observations of dyes injected into liquids led Reynolds [121], in 1883, to conclude that the orderliness of the flow was dependent on its velocity. At small velocities the flow appeared orderly, in layers, that is, *laminar*. With increasing velocities, Reynolds observed a mixing between the dye and water; the pattern of flow became irregular, or *turbulent*.

Within the range of laminar flow, Reynolds found a linear proportionality to exist between the hydraulic gradient and the velocity of flow, in keeping with Darcy's law. With the advent of turbulence, the hydraulic gradient approached the square of the velocity. These observations suggest a representation of the hydraulic gradient as

$$i = av + bv^n \tag{1}$$

where a and b are constants and n is between 1 and 2. From studies of the flow of water through columns of shot of uniform size, Lindquist [99] reports n to be exactly 2. Whatever the precise order of n, experiments have shown conclusively that for small velocities (within the laminar range), Darcy's law gives an accurate representation of the flow within a porous medium.

There remains now the question of the determination of the laminar range of flow and the extent to which actual flow systems through soils are included. Such a criterion is furnished by Reynolds number R (a pure number relating inertial to viscous force), defined as

$$R = \frac{vD\rho}{\mu} \tag{2}$$

where v = discharge velocity, cm/sec
$\quad\quad D$ = average of diameters of soil particles, cm
$\quad\quad \rho$ = density of fluid, g (mass)/cm^3
$\quad\quad \mu$ = coefficient of viscosity, g-sec/cm^2

The critical value of Reynolds number at which the flow in soils changes from laminar to turbulent flow has been found by various investigators [99] to range between 1 and 12. However, it will suffice in all our work to accept the validity of Darcy's law when Reynolds number is taken as equal to or less than unity, or

$$\frac{vD\rho}{\mu} \leqq 1 \tag{3}$$

Substituting the known values of ρ and μ for water into Eq. (3) and assuming a conservative velocity of $\frac{1}{4}$ cm/sec, we have D equal to 0.4 mm, which is representative of the average particle size of coarse sand. Hence it can be concluded that it is most unlikely that the range of validity of Darcy's law will be exceeded in natural flow situations. It is interesting to note that the laminar character of flow encountered in soils represents one of few valid examples of such flow in all hydraulic engineering. An excellent discussion of the "law of flow" and a summary of investigations are to be found in Muskat [99].

1-6. Coefficient of Permeability

For laminar flow we found that Darcy's law could be written as [Eq. (4), Sec. 1-4]

$$v = -k\frac{dh}{ds} \tag{1}$$

where k is the coefficient of permeability.

Owing to the failure at the present time to attain more than a moderately valid expression relating the coefficient of permeability to the geometric characteristics of soils,* refinements in k are hardly warranted. However, because of the influence that the density and viscosity of the pore fluid may exert on the resulting velocity, it is of some value to isolate

---

* Notable among these are the Kozeny-Carman equation [13, 75], and more recently the equation of Schmid [128]. An excellent discussion of their shortcomings is given by Leonards [87].

that part of k which is dependent on these properties. To do this, we introduce the *physical permeability* k_0 (square centimeters), which is a constant typifying the structural characteristics of the medium and is independent of the properties of the fluid. The relationship between the *permeability* and the *coefficient of permeability* as given by Muskat [99] is

$$k = k_0 \frac{\gamma_w}{\mu} \tag{2}$$

where γ_w is the unit weight of the fluid and μ is the coefficient of viscosity.

Substituting Eq. (2) into Darcy's law, we obtain

$$v = -k_0 \frac{\gamma_w}{\mu} \frac{dh}{ds} \tag{3}$$

which indicates that the discharge velocity is inversely proportional to the viscosity of the fluid. Equation (3) may be used when dealing with more than one fluid or with temperature variations. In the groundwater and seepage problems encountered in civil engineering, where we are primarily interested in the flow of a single relatively incompressible fluid subject to small changes in temperature, it is more convenient to use Darcy's law with k as in Eq. (1).

Laboratory and field determinations of k have received such excellent coverage in the soil-mechanics literature [81, 99, 142, 145] that duplication in this book does not appear to be warranted. Some typical values of the coefficient of permeability are given in Table 1-2.

Table 1-2. Some Typical Values of Coefficient of Permeability*

Soil type	Coefficient of permeability k, cm/sec
Clean gravel	1.0 and greater
Clean sand (coarse)	1.0–0.01
Sand (mixture)	0.01–0.005
Fine sand	0.05–0.001
Silty sand	0.002–0.0001
Silt	0.0005–0.00001
Clay	0.000001 and smaller

* From A. I. Silin-Bekchurin, "Dynamics of Underground Water," p. 34, Moscow University, 1958.

1-7. Capillarity

Although the physical cause of capillarity is subject to controversy, the surface-tension concept of capillarity renders it completely amenable to rational analysis.

If a tube filled with dry sand has its lower end immersed below the level of free water (Fig. 1-6a), water will rise within the sand column to an elevation h_c called the *height of capillary rise*. Consideration of the

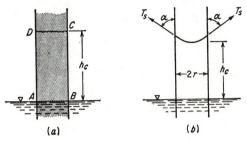

FIG. 1-6

equilibrium of a column of water in a capillary tube of radius r (Fig. 1-6b) yields, for h_c,

$$h_c = \frac{2T_s}{r\gamma_w} \cos \alpha \tag{1}$$

T_s is the *surface tension* of the water (≈ 0.075 g/cm). It can be thought of as being analogous to the tension in a membrane acting at the air-water interface (meniscus) and supporting a column of water of height h_c. α is called the *contact angle* and is dependent on the chemical properties of both the tube and the water. For clean glass and pure water, $\alpha = 0$.

If atmospheric pressure is designated as p_a, then the pressure in the water immediately under the meniscus will be

$$p = p_a - \gamma_w h_c \tag{2}$$

On the assumption that the rate of capillary rise is governed by Darcy's law, Terzaghi [144] obtained from the differential equation

$$n\frac{dy}{dt} = k\frac{h_c - y}{y}$$

the following expression for the time t required for the capillary water to rise to the height y:

$$t = \frac{nh_c}{k}\left[\ln\left(\frac{h_c}{h_c - y}\right) - \frac{y}{h_c}\right] \tag{3}$$

As the capillary water rises, air becomes entrapped in the larger pores and hence variations are induced in both h_c and k. From laboratory tests, Taylor [142] found that for $y/h_c < 20$ per cent the degree of saturation is relatively high and Eq. (3) can be considered valid.

On the basis of Eqs. (1) and (3) we see that although the fine-grained soils exhibit a greater capillary potential than the granular soils their minimal coefficients of permeability greatly reduce their rates of capillary rise. Some typical values for the height of capillary rise are presented in Table 1-3.

Table 1-3. Typical Values of Height of Capillary Rise*

Soil type	Height of capillary rise h_c, cm
Coarse sand	2–5
Sand	12–35
Fine sand	35–70
Silt	70–150
Clay	200–400 and greater

* From A. I. Silin-Bekchurin, "Dynamics of Underground Water," p. 23, Moscow University, 1958.

1-8. General Hydrodynamic Equations, Velocity Potential

In Fig. 1-7, $\bar{u}$, $\bar{v}$, and $\bar{w}$ are the components of the seepage velocity at the point in the fluid $A(x,y,z)$ at the time t. Thus they represent func-

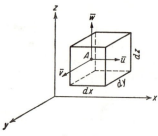

tions of the independent variables x, y, z, and t. For a particular value of t, they specify the motion at all points occupied by the fluid; and for any point within the fluid, they are functions of time, giving a history of the variations of velocity at that point. Unless otherwise specified, $\bar{u}$, $\bar{v}$, $\bar{w}$, and their space derivatives ($\partial \bar{u}/\partial x$, $\partial \bar{v}/\partial y$, $\partial \bar{w}/\partial z$) are everywhere bounded.

Fig. 1-7

A particle of fluid originally at point $A(x,y,z)$ at time t will move during the time δt to the position $(x + \bar{u}\delta t, y + \bar{v}\delta t, z + \bar{w}\delta t)$. The time rate of change of any velocity component, say $\delta \bar{u}/\delta t$, will be

$$\frac{\delta \bar{u}}{\delta t} = \frac{\partial \bar{u}}{\partial t} + \frac{\partial \bar{u}}{\partial x}\frac{\delta x}{\delta t} + \frac{\partial \bar{u}}{\partial y}\frac{\delta y}{\delta t} + \frac{\partial \bar{u}}{\partial z}\frac{\delta z}{\delta t}$$

Now if δt is considered to approach zero, noting that $\bar{u} = \dfrac{dx}{dt}$, $\bar{v} = \dfrac{dy}{dt}$, and $\bar{w} = \dfrac{dz}{dt}$, the total acceleration in the x direction is

$$\frac{d\bar{u}}{dt} = \frac{\partial \bar{u}}{\partial t} + \bar{u}\frac{\partial \bar{u}}{\partial x} + \bar{v}\frac{\partial \bar{u}}{\partial y} + \bar{w}\frac{\partial \bar{u}}{\partial z} \tag{1a}$$

Similarly, in the y and z directions,

$$\frac{d\bar{v}}{dt} = \frac{\partial \bar{v}}{\partial t} + \bar{u}\frac{\partial \bar{v}}{\partial x} + \bar{v}\frac{\partial \bar{v}}{\partial y} + \bar{w}\frac{\partial \bar{v}}{\partial z} \tag{1b}$$

$$\frac{d\bar{w}}{dt} = \frac{\partial \bar{w}}{\partial t} + \bar{u}\frac{\partial \bar{w}}{\partial x} + \bar{v}\frac{\partial \bar{w}}{\partial y} + \bar{w}\frac{\partial \bar{w}}{\partial z} \tag{1c}$$

Let the pressure of the center point $A(x,y,z)$ of Fig. 1-7 be p and the density of the fluid be ρ and let the components of the body force per unit mass be, respectively, X, Y, and Z in the x, y, and z directions at the time t. In groundwater flow the common body force is that due to the force of gravity. With a pressure p at point $A(x,y,z)$, the force on the yz face of the element nearest the origin is

$$\left(p - \frac{1}{2} \frac{\partial p}{\partial x} dx \right) dy \, dz$$

and that on the opposite face is

$$\left(p + \frac{1}{2} \frac{\partial p}{\partial x} dx \right) dy \, dz$$

By Newton's second law of motion, the product of the mass and acceleration in the x direction must equal the sum of the forces in that direction; hence

$$\rho \, dx \, dy \, dz \frac{d\bar{u}}{dt} = \left(p - \frac{1}{2} \frac{\partial p}{\partial x} dx \right) dy \, dz - \left(p + \frac{1}{2} \frac{\partial p}{\partial x} dx \right) dy \, dz + \rho X \, dx \, dy \, dz$$

Simplifying and substituting the value of the total acceleration from Eq. (1a), we have

$$\frac{\partial \bar{u}}{\partial t} + \bar{u} \frac{\partial \bar{u}}{\partial x} + \bar{v} \frac{\partial \bar{u}}{\partial y} + \bar{w} \frac{\partial \bar{u}}{\partial z} = X - \frac{1}{\rho} \frac{\partial p}{\partial x} \qquad (2a)$$

Similarly, in the y and z directions,

$$\frac{\partial \bar{v}}{\partial t} + \bar{u} \frac{\partial \bar{v}}{\partial x} + \bar{v} \frac{\partial \bar{v}}{\partial y} + \bar{w} \frac{\partial \bar{v}}{\partial z} = Y - \frac{1}{\rho} \frac{\partial p}{\partial y} \qquad (2b)$$

$$\frac{\partial \bar{w}}{\partial t} + \bar{u} \frac{\partial \bar{w}}{\partial x} + \bar{v} \frac{\partial \bar{w}}{\partial y} + \bar{w} \frac{\partial \bar{w}}{\partial z} = Z - \frac{1}{\rho} \frac{\partial p}{\partial z} - g \qquad (2c)$$

Equations (2) are Euler's equations [80] of motion for a nonviscous fluid.

For laminar flow, the components of velocity and their space derivatives are known to be small; hence the product $\bar{u} \dfrac{\partial \bar{u}}{\partial x}$, etc., can be neglected and Eqs. (2) become (velocity component without bar is discharge velocity)

$$\frac{1}{n} \frac{\partial u}{\partial t} = X - \frac{1}{\rho} \frac{\partial p}{\partial x} \qquad (3a)^{*}$$

$$\frac{1}{n} \frac{\partial v}{\partial t} = Y - \frac{1}{\rho} \frac{\partial p}{\partial y} \qquad (3b)$$

$$\frac{1}{n} \frac{\partial w}{\partial t} = Z - \frac{1}{\rho} \frac{\partial p}{\partial z} - g \qquad (3c)$$

* The development from Eqs. (3) to Eqs. (7) was first given by Risenkampf [122].

When the state of flow is independent of the time (steady state), the left sides of Eqs. (3) vanish. For example, Eq. (3a) becomes

$$X = \frac{1}{\rho}\frac{\partial p}{\partial x} \tag{4}$$

Hence, substituting for p from Eq. (3), Sec. 1-4, Eq. (4) takes the form (ρ is constant)

$$X = g\frac{\partial h}{\partial x} = -gi \tag{5}$$

where i is the hydraulic gradient. Applying Darcy's law, we have finally

$$X = -\frac{gu}{k} \tag{6a}$$

Similarly, for Y and Z we obtain

$$Y = -\frac{gv}{k} \tag{6b}$$

$$Z = -\frac{gw}{k} \tag{6c}$$

Equations (6) indicate that for steady-state and laminar flow the body forces are linear functions of the velocity.

Assuming that the coefficient of permeability k is independent of the state of flow and substituting Eqs. (6) into Eqs. (3), we obtain the following dynamical equations of flow:

$$\begin{aligned}
\frac{1}{n}\frac{\partial u}{\partial t} &= -\frac{1}{\rho}\frac{\partial p}{\partial x} - \frac{gu}{k} \\
\frac{1}{n}\frac{\partial v}{\partial t} &= -\frac{1}{\rho}\frac{\partial p}{\partial y} - \frac{gv}{k} \\
\frac{1}{n}\frac{\partial w}{\partial t} &= -\frac{1}{\rho}\frac{\partial p}{\partial z} - \frac{gw}{k} - g
\end{aligned} \tag{7}$$

For steady flow Eqs. (7) reduce to the vectorial generalization of Darcy's law,

$$\mathbf{v} = -k\operatorname{grad} h \tag{8a}$$

Although the steady-state assumption is generally made to set aside the inertia terms in Eqs. (7) [88, 99], it can be shown that these terms are of negligible order in a wide range of situations, even when velocity changes do occur with time. To demonstrate this, we note that each of Eqs. (7) can be expressed in the form (i is hydraulic gradient)

$$\frac{1}{n}\frac{\partial v}{\partial t} = gi - \frac{gv}{k}$$

which, with the change of variable

$$v = v_* + ki \tag{8b}$$

reduces to the expression

$$\frac{1}{n}\frac{\partial v_*}{\partial t} = -\frac{k}{n}\frac{\partial i}{\partial t} - \frac{g v_*}{k}$$

Substituting conservative values for n, g, and k [$n = \frac{1}{2}$, $g \approx 1000$ cm/sec² and $k = 0.1$ cm/sec (coarse sand)], we obtain

$$\frac{\partial v_*}{\partial t} = -1 \times 10^{-1}\frac{\partial i}{\partial t} - \frac{1}{2} \times 10^4 v_* \qquad (8c)$$

Comparing the orders of magnitude of the two terms on the right side of this expression, we see that if the ratio $(\partial i/\partial t)/v_*$ is not exceedingly large, the first right-hand term is a small quantity and can be neglected. Thus, the remaining equation is

$$\frac{dv_*}{dt} = -\frac{1}{2} \times 10^4 v_*$$

which, after integration, yields

$$v_* = v_{*_0} \exp\left(-\frac{1}{2} \times 10^4 t\right)$$

The right part of this equation very rapidly tends to zero and hence within a fraction of a second we can consider $v_* = 0$ and Eq. (8b) will reduce to Eq. (8a). Analytical studies conducted by Pavlovsky [110] and Davison [26] indicate that unless reservoir water elevations are subject to excessive variations in time the contribution of $\partial i/\partial t$ in Eq. (8c) can be neglected. Physically, this implies that the speed of flow through natural soils for laminar flow is so slow that changes in momentum are negligible in comparison with the viscous resistance to flow. In some measure, the nominal contribution of the inertia terms was implied previously in the critical value of Reynolds number which assured laminar flow and the validity of Darcy's law. If one recalls (Sec. 1-5) that Reynolds number is the ratio of inertia force to viscous force and that for laminar flow in soils $R \leqq 1$, it is evident that the viscous forces are at least of the order of magnitude of the inertia forces. This is in contradistinction to considerations of laminar flow through pipes ($R_{critical} = 2,000$) where the forces due to momentum changes may be very much greater than the viscous forces resisting the flow.

Equation (8a) contains the four unknowns u, v, w, and h. Hence one more equation must be added to make the system complete. This is the *equation of continuity* which assumes that the fluid is continuous in space and time.

The quantity of fluid through the yz face of the element of Fig. 1-7 nearest the origin is

$$n\bar{u}\,dy\,dz$$

and that through the opposite face is

$$\left[n\bar{u} + \frac{\partial}{\partial x} (n\bar{u}) \, dx \right] dy \, dz$$

The net gain in the quantity of fluid per unit time in the x direction is

$$\frac{\partial}{\partial x} (n\bar{u}) \, dx \, dy \, dz$$

Similarly, the gains in the y and z directions are, respectively,

$$\frac{\partial}{\partial y} (n\bar{v}) \, dx \, dy \, dz$$

$$\frac{\partial}{\partial z} (n\bar{w}) \, dx \, dy \, dz$$

If the fluid and flow medium are both incompressible, the total gain of fluid per unit time must be identically zero; hence

$$\frac{\partial}{\partial x} (n\bar{u}) + \frac{\partial}{\partial y} (n\bar{v}) + \frac{\partial}{\partial z} (n\bar{w}) = 0 \tag{9}$$

and

$$\frac{\partial u}{\partial x} + \frac{\partial v}{\partial y} + \frac{\partial w}{\partial z} = 0 \tag{10}$$

Equation (10) is the *equation of continuity in three dimensions.*

It is of the utmost convenience in groundwater flow to introduce the *velocity potential ϕ*, defined as

$$\phi(x,y,z) = -k \left(\frac{p}{\gamma_w} + z \right) + C = -kh + C \tag{11}*$$

where C is an arbitrary constant. Thus,

$$u = \frac{\partial \phi}{\partial x} \qquad v = \frac{\partial \phi}{\partial y} \qquad w = \frac{\partial \phi}{\partial z} \tag{12}$$

Equations (11) and (12) represent the *generalized Darcy's law* which provide the dynamical framework for all investigations into groundwater flow.

Substituting Eqs. (12) into the equation of continuity [Eq. (10)], we obtain the *Laplace equation* (see Sec. A-3)

$$\nabla^2 \phi = \frac{\partial^2 \phi}{\partial x^2} + \frac{\partial^2 \phi}{\partial y^2} + \frac{\partial^2 \phi}{\partial z^2} = 0 \tag{13}$$

---

* The concept of the velocity potential ϕ will become apparent in the development that follows. Analogous to a force potential whose directional derivative is the force in that direction, the velocity potential is a scalar function of space such that its derivative with respect to any direction is the velocity of the fluid in that direction.

Equation (13) indicates that for conditions of steady-state, laminar flow, the form of the groundwater motion can be completely determined by solving one equation, subject to the boundary conditions of the flow domain.

1-9. Two-dimensional Flow, Stream Function

Physically, all flow systems extend in three dimensions. However, in many problems the features of the groundwater motion are essentially planar, with the motion being substantially the same in parallel planes. For these problems we need concern ourselves with two-dimensional flow only, and thereby we are able to reduce considerably the work necessary to effect a solution. Fortunately, in civil engineering the vast majority of problems falls into this category.

The fundamental equations for two-dimensional flow in the xy plane will be taken as

$$u = \frac{\partial \phi}{\partial x} = -k \frac{\partial h}{\partial x} \qquad v = \frac{\partial \phi}{\partial y} = -k \frac{\partial h}{\partial y} \tag{1}$$

Assuming the y axis as vertical (positive up) and the x axis as horizontal,

$$h = \frac{p}{\gamma_w} + y \qquad \phi = -k \left(\frac{p}{\gamma_w} + y \right) + C \tag{2}$$

Correspondingly, Laplace's equation reduces to

$$\nabla^2 \phi = \frac{\partial^2 \phi}{\partial x^2} + \frac{\partial^2 \phi}{\partial y^2} = 0 \tag{3}$$

and the equation of continuity becomes

$$\frac{\partial u}{\partial x} + \frac{\partial v}{\partial y} = 0 \tag{4}$$

In Appendix A are discussed some of the general properties of Laplace's equation. In particular, it is shown that Laplace's equation is satisfied by the conjugate harmonic functions ϕ and ψ and that the curves $\phi(x,y) = $ constant are the orthogonal trajectories of the curves $\psi(x,y) = $ constant. In groundwater flow literature the function $\psi(x,y)$ is called the *stream function* and is defined as

$$u = \frac{\partial \psi}{\partial y} \qquad v = -\frac{\partial \psi}{\partial x} \tag{5}*$$

Substituting Eqs. (5) into the equation of continuity [Eq. (4)], we obtain

$$\frac{\partial^2 \psi}{\partial x\, \partial y} - \frac{\partial^2 \psi}{\partial y\, \partial x} = 0$$

---

* For simplicity, $\psi(x,y)$ and $\phi(x,y)$ are generally written as ψ and ϕ; however, their dependence on both x and y is implied.

Equating the respective potential and stream functions of u and v

$$\frac{\partial \phi}{\partial x} = \frac{\partial \psi}{\partial y} \qquad \frac{\partial \phi}{\partial y} = -\frac{\partial \psi}{\partial x} \tag{6}$$

we see that $\psi(x,y)$ satisfies identically the equation of continuity and the Cauchy-Riemann equations and hence the equation of Laplace,

$$\nabla^2 \psi = \frac{\partial^2 \psi}{\partial x^2} + \frac{\partial^2 \psi}{\partial y^2} = 0 \tag{7}$$

1-10. Streamlines and Equipotential Lines

Consider AB of Fig. 1-8 as the path of a particle of fluid passing through point $P(x,y)$ with a tangential velocity V. We see immediately from the figure that

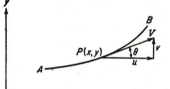

$$\frac{v}{u} = \tan \theta = \frac{dy}{dx}$$

and hence

FIG. 1-8

$$v \, dx - u \, dy = 0 \tag{1}$$

Substituting Eqs. (5) of Sec. 1-9 into Eq. (1), it follows that

$$\frac{\partial \psi}{\partial x} dx + \frac{\partial \psi}{\partial y} dy = 0$$

and therefore the total differential $d\psi = 0$ and

$$\psi = \text{constant} \tag{2}$$

Thus we see that the curves $\psi(x,y)$, equal to a sequence of constants, are at all points tangent to the velocity vectors and hence define the path of flow. The locus of the path of flow of an individual particle of water is called a *flow line* or *streamline*.

Another important physical property of the stream function can be obtained by considering the flow between the two streamlines ψ_1 and ψ_2 of Fig. 1-9. If the quantity of discharge through the line ab is q, then

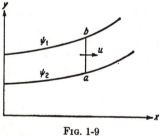

$$q = \int_{\psi_2}^{\psi_1} u \, dy = \int_{\psi_2}^{\psi_1} d\psi = \psi_1 - \psi_2 \tag{3}*$$

FIG. 1-9

Equation (3) presents an important flow characteristic which will be used

* The symbol q will be used to denote the quantity of discharge per unit length normal to the plane of flow (cubic feet per second per foot). The symbol Q will be reserved for the volume discharge (cubic feet per second).

often in subsequent work; namely, that the quantity of flow between two streamlines, called a *flow channel*, is a constant. Thus, once the streamlines of flow have been obtained, their plot not only shows the direction of flow but the relative magnitudes of the velocity along the flow channels, i.e., the velocity at any point in the flow channel varies inversely with the streamline spacing in the vicinity of that point.

The physical significance of ϕ can be obtained from consideration of the total differential along the curve $\phi(x,y) = $ constant,

$$d\phi = \frac{\partial \phi}{\partial x} dx + \frac{\partial \phi}{\partial y} dy = 0$$

Substituting from Eq. (1), Sec. 1-9, for $\partial \phi / \partial x$ and $\partial \phi / \partial y$, we have

$$u\, dx + v\, dy = 0$$

and
$$\frac{dy}{dx} = -\frac{u}{v} \tag{4}$$

The curves $\phi(x,y) = C$, where C is a sequence of constants, are called *equipotential lines;* as seen from Eqs. (1) and (4) the streamlines and equipotential lines form a grid of curves called a *flow net* wherein all intersections are at right angles.

An important distinction between the ϕ and ψ functions lies in the fact that the ϕ functions exist only for irrotational flow. A particle of fluid is said to have zero net rotation or to be irrotational if the circulation, the line integral of the tangential velocity taken around the particle, is zero. This may be visualized by considering the particle as a free body of fluid in the shape of a sphere. As the fluid may be considered frictionless (Sec. 1-4), all surface forces act normal to the surface and hence through its mass center. Likewise, gravity acts through the mass center, which, for an incompressible fluid, is coincidental with its geometric center. Thus, no torque can exist on the sphere and it remains without rotation. This principle can best be illustrated

FIG. 1-10

by example. In Fig. 1-10, $ABCD$ represents a rectangular element in two-dimensional flow. The circulation in this case is

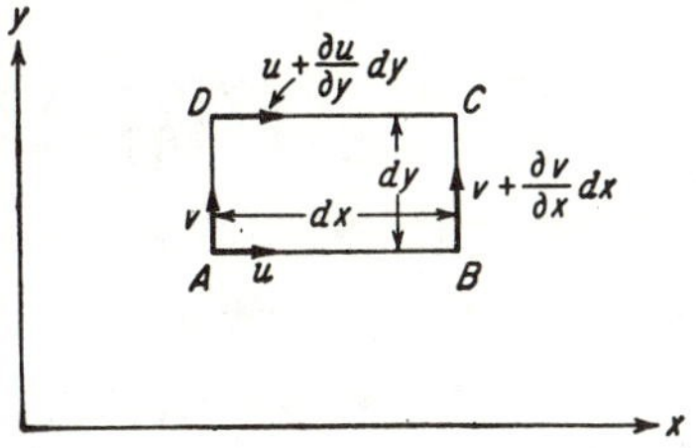

$$u\, dx + \left(v + \frac{\partial v}{\partial x} dx \right) dy - \left(u + \frac{\partial u}{\partial y} dy \right) dx - v\, dy$$

Now if the circulation is zero, we have for irrotational flow,

$$\frac{\partial v}{\partial x} - \frac{\partial u}{\partial y} = 0 \tag{5}$$

Substituting for u and v, we find that

$$\frac{\partial^2 \phi}{\partial x\, \partial y} - \frac{\partial^2 \phi}{\partial x\, \partial y} = 0$$

which shows that the existence of the velocity potential implies that flow is irrotational.

It should be noted that within a given region of flow the streamlines and equipotential lines are unique. That is, considering the total differential

$$d\psi = \frac{\partial \psi}{\partial x}\, dx + \frac{\partial \psi}{\partial y}\, dy$$

we see immediately from the Cauchy-Riemann equations [Eqs. (6), Sec. 1-9] that

$$\psi = \int \left(\frac{\partial \phi}{\partial x}\, dy - \frac{\partial \phi}{\partial y}\, dx \right) \tag{6a}$$

and

$$\phi = \int \left(\frac{\partial \psi}{\partial y}\, dx - \frac{\partial \psi}{\partial x}\, dy \right) \tag{6b}$$

Hence in solving groundwater problems we need concern ourselves only with the determination of one of the functions, subject to the imposed boundary conditions. The other function will follow directly from Eqs. (6).

A combination of the functions ϕ and ψ, called the *complex potential* and defined by

$$w = \phi + i\psi \tag{7}$$

is of particular interest, as $\nabla^2 w = \nabla^2 \phi + i\nabla^2 \psi = 0$ satisfies both Eqs. (3) and (7), Sec. 1-9, and hence insures both irrotational and continuous flow.

1-11. Boundary Conditions

In the general case of plane, steady-state flow of groundwater through homogeneous soils, four types of boundaries are encountered. The characteristics of each of these boundaries will be considered in detail.

1. *Impervious boundary.* At impervious boundaries the fluid can neither penetrate the boundary nor leave gaps; thus the velocity component normal to the boundary at any point must vanish. As the fluid is assumed to be frictionless, no restrictions are placed on its tangential components. Defining n and t as the normal and tangential directions, respectively, at a point on the boundary, from Eqs. (6), Sec. 1-9, we have

$$\frac{\partial \phi}{\partial n} = \frac{\partial \psi}{\partial t} = 0 \qquad \psi = \text{constant}$$

Hence an impervious boundary is seen to define the locus of a streamline. Likewise, any streamline satisfies the condition for an impervious boundary and may be taken as such.   Two types of impervious boundaries are shown in Fig. 1-11.   One, AB of the figure, is generally the upper surface

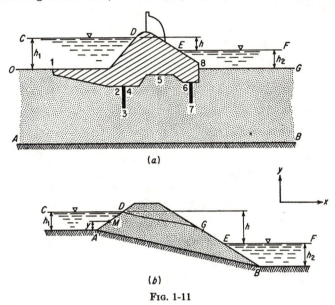

FIG. 1-11

of a soil stratum or rock wherein the coefficient of permeability is insignificant in comparison to that of the soil above.   In this case the impervious boundary defines the lowest streamline.   The other, representing the upper flow line, is the bottom contour of the impervious structure (1-8 in Fig. 1-11a).

2. *Boundaries of the reservoirs.*   Along the boundaries of the reservoir the pressure distribution may be taken as hydrostatic.   Therefore, at any point such as M along the boundary AD of Fig. 1-11b, the pressure in the water is

$$p = \gamma_w(h_1 - y) \qquad (1)$$

Eliminating the common pressure terms from Eq. (1) and Eq. (2) of Sec. 1-9,

$$\phi = -kh_1 + C \qquad (2)$$

Since k, C, and h_1 are all constants,

$$\phi = \text{constant} \qquad (3)$$

and thus all reservoir boundaries, such as $O1$ and $8G$ of Fig. 1-11a and AD and EB of Fig. 1-11b, are equipotential lines.

3. *Surface of seepage.* The surface of seepage (*GE* of Fig. 1-11*b*) represents a boundary where the seepage leaving the flow region enters a zone free of both liquid and soil. As the pressure on this surface is both constant and atmospheric, and since the surface is neither an equipotential line nor a streamline, along this boundary

$$\phi = -\frac{kp}{\gamma_w} - ky + C$$

hence we obtain the linear relationship

$$\phi + ky = \text{constant} \tag{4}$$

4. *Line of seepage* (*free surface, depression curve*). The line of seepage is the upper streamline in the flow domain. It separates the saturated region of flow from that part of the soil body through which no flow occurs, such as *DG* of Fig. 1-11*b*. The determination of its locus is one of the major objectives of groundwater investigations. In addition to the requirement that the line of seepage be a streamline ($\psi = \text{constant}$) it is evident that the pressure at every point along its surface is constant and equal to atmospheric pressure. Thus, along this line,

$$\phi + ky = \text{constant} \tag{5*}$$

which demonstrates that the velocity potential (and total head) along the line of seepage varies linearly with elevation head. This requires

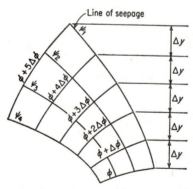

Fig. 1-12

constant vertical intercepts ($\Delta y = \text{constant}$) at the points of intersection of the line of seepage with successive equipotential lines of equal drops ($\Delta\phi$ of Fig. 1-12).

* Although both the line of seepage and the surface of seepage must satisfy the same equation [Eqs. (4) and (5)], the line of seepage is a streamline whereas the surface of seepage is not.

Various entrance and emergence conditions for the line of seepage are given in Fig. 1-13. (For the derivation and experimental verification of these, see Refs. 14 and 16.)

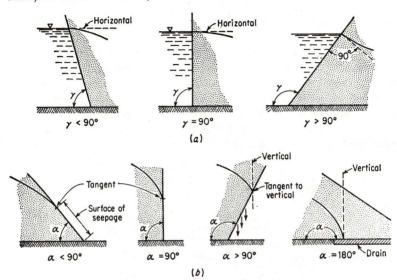

FIG. 1-13. *(After A. Casagrande [14].)*

1-12. The Flow Net

A graphical representation of the family of streamlines and their corresponding equipotential lines within a flow region is called a *flow net*. The orthogonal network in Fig. 1-15 represents such a system. Although the graphical construction of a flow net often requires tedious trial-and-error adjustments, it is one of the more valuable methods employed in two-dimensional flow problems.

If, in Fig. 1-14, Δn denotes the distance between a pair of adjacent streamlines and Δs is the distance between a pair of adjacent equipotential lines at some point A within the region of flow, the approximate velocity at point A will be

$$v_A \approx \frac{\Delta \phi}{\Delta s} \approx \frac{\Delta \psi}{\Delta n} \qquad (1)$$

FIG. 1-14

Since the quantity of flow between any two streamlines (Δq) is a constant and equal to $\Delta \psi$, we have from Eq. (1)

$$\Delta q \approx \frac{\Delta n}{\Delta s} \Delta \phi \qquad (2)$$

It is expedient at this point to introduce into Eq. (2) the *reduced quantity of seepage* and the total head

$$\bar{q} = \frac{q}{k} \qquad h = \frac{\phi}{k}$$

which result in the expression

$$\Delta\bar{q} \approx \frac{\Delta n}{\Delta s} \Delta h \tag{3}$$

The above expressions are approximate when Δn and Δs are finite; however, as these distances become very small, Eq. (1) approaches the correct value for the velocity at point A, and Eq. (3) yields the exact value for the reduced quantity of seepage through the flow channel.

In Fig. 1-15 there are four known boundaries: two streamlines [the bottom contour of the structure ($\bar{q} = 0$) and the surface of the impervious

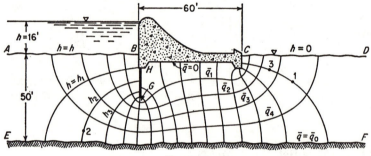

Fig. 1-15

layer EF ($\bar{q} = \bar{q}_0$)], and two equipotential lines [the boundaries at the reservoirs AB ($h = h$) and CD ($h = 0$)]. The intermediate streamlines (such as $\bar{q} = \bar{q}_1, \bar{q}_2, \bar{q}_3, \ldots$) and equipotential lines ($h = h_1, h_2, h_3, \ldots$) must intersect each other and the reservoir boundaries at right angles. If, from the infinite number of streamlines and equipotential lines, we specify the same drop in head (Δh) between adjacent equipotential lines and the same reduced quantity of seepage ($\Delta\bar{q}$) between neighboring flow lines, according to Eq. (3), in the limit, the resulting flow net will be composed entirely of rectangles with the same $\Delta n/\Delta s$ ratio.

A graphical technique of constructing flow nets, based upon the converse of the foregoing, was first suggested by Prášil [118], although it was developed formally by Forchheimer [37]. If one plots the nests of streamlines and equipotential lines so that they preserve right-angle intersections, satisfy the boundary conditions, and form curvilinear squares*

---

* In speaking of squares, we except *singular* squares such as the *five-sided square* at H in Fig. 1-15 and the *three-sided square* at G. However, when they are subdivided into smaller squares, it is immediately apparent that these deviations reduce in size and, in the limit, act only at singular *points*, the effects of which may be disregarded.

($\Delta n/\Delta s = 1$ is most sensitive to visual inspection) which reduce to perfect squares in the limit as the number of lines is increased, then one has obtained an unique solution of Laplace's equation for the flow region from which the quantity of seepage, seepage pressures, etc., can be had easily. For example, designating N_f as the number of flow channels and N_e as the number of equipotential drops along each of the channels, we have immediately from Eq. (3) (with $\Delta n/\Delta s = 1$) for the quantity of seepage

$$q = N_f k \,\Delta\bar{q} = \frac{N_f}{N_e} kh \tag{4}$$

where $h = N_e \,\Delta h$ is the total loss in head. In Fig. 1-15 we see that N_f equals about 5 and N_e equals 16.

The following procedure is suggested for the construction of a flow net:

1. Draw the boundaries of the flow region to scale so that all equipotential lines and streamlines that are drawn can be terminated on these boundaries.

2. Sketch lightly three or four streamlines, keeping in mind that they are only a few of the infinite number of curves that must provide a smooth transition between the boundary streamlines. As an aid in the spacing of these lines, it should be noted that the distance between adjacent streamlines increases in the direction of the larger radius of curvature.

3. Sketch the equipotential lines, bearing in mind that they must intersect all streamlines, including the boundary streamlines, at right angles and that the enclosed figures must be squares.*

4. Adjust the locations of the streamlines and the equipotential lines to satisfy the requirements of step 3. This is a trial-and-error process with the amount of correction being dependent upon the position of the initial streamlines. The speed with which a successful flow net can be drawn is highly contingent on the experience and judgement of the individual. In this regard, the beginner will find the suggestions in A. Casagrande's paper [14] to be of particular assistance.

5. As a final check on the accuracy of the flow net, draw the diagonals of the squares. These should also form smooth curves which intersect each other at right angles.

1-13. Seepage Force and Critical Gradient

By virtue of the viscous friction exerted on water flowing through the soil pores, an energy transfer is effected between the water and the soil. The measure of this transfer we found to be the head loss (Δh of Fig. 1-5) between the points under consideration (Δs). The force corresponding to this energy transfer is called the *seepage force*. It is this seepage force

---

* See previous footnote.

that is responsible for the phenomenon known as quicksand [144] and is of vital importance in the stability analyses of earth structures subject to the action of seepage.

The first rational approach to the problem was presented by Terzaghi in 1922 [143] and forms the basis of all subsequent studies. The theory will be presented in a somewhat modified form.

Let us consider all the forces acting on a unit volume of soil through which seepage occurs.

1. The weight of solids per unit volume γ_0 is

$$\gamma_0 = \frac{S_s \gamma_w}{1 + e}$$

where e is the void ratio, S_s is the specific gravity, and γ_w is the unit weight of water. If we define a unit vertical vector $\mathbf{j}$, positive up, we have for γ_0

$$\gamma_0 = -\mathbf{j} \frac{S_s \gamma_w}{1 + e} \tag{1}$$

2. The second force γ_1 is the weight of water per unit volume displaced by the solid particles

$$\gamma_1 = -\mathbf{j} \frac{\gamma_w}{1 + e} \tag{2}$$

The difference between γ_0 and γ_1,

$$\gamma'_m = -\mathbf{j} \frac{\gamma_w(S_s - 1)}{1 + e} \tag{3}$$

is called the *submerged unit weight*.

3. The total weight per unit volume of the soil-water mass γ_m is

$$\gamma_m = -\mathbf{j} \frac{\gamma_w(S_s + e)}{1 + e} \tag{4}$$

The difference between Eqs. (3) and (4) is simply

$$-\mathbf{j}\gamma_w \tag{5}$$

These results are plotted on the vertical axis of Fig. 1-16 and represent the hydrostatic forces per unit volume acting within the flow medium. To account for the hydrodynamic forces, Eq. (2), Sec. 1-9, is written as

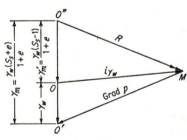

FIG. 1-16

$$\frac{p}{\gamma_w} = h - y$$

Taking the gradient of both sides of this equation we obtain

$$\frac{1}{\gamma_w} \text{ grad } p = \text{grad } h - \mathbf{j} \qquad (6)$$

where $\mathbf{j}$ is a unit vector, as before.   Multiplying Eq. (6) by γ_w and replacing grad h by $-i$, the hydraulic gradient, we have the vector equation

$$\text{grad } p = -i\gamma_w - \mathbf{j}\gamma_w \qquad (7)$$

Equation (7) is plotted as triangle $OO'M$ in Fig. 1-16.* $i\gamma_w(OM)$ represents the seepage force per unit volume, the direction of which is normal to the equipotentials; $R(O''M)$ represents the magnitude and direction of the resultant force (per unit volume) acting within the pore water at a point in the soil.

For $R = 0$, we see immediately from Fig. 1-16 that a quick condition is incipient if

$$i_{cr} = \frac{S_s - 1}{1 + e} = \frac{\gamma_m'}{\gamma_w} \qquad (8)$$

Substituting typical values of $S_s = 2.65$ (quartz sand) and $e = 0.65$ (for sand, $0.57 \leq e \leq 0.95$) we see that as an average value the critical gradient can be taken as

$$i_{cr} \approx 1 \qquad (9)$$

When information is lacking as to the specific gravity and void ratio of the soil, the critical gradient is generally taken as unity [Eq. (9)].

Equations (8) and (9) provide the basis for stability determinations of the factor of safety against a quick condition (called *piping*).   In essence the procedure requires the determination of the maximum hydraulic gradient along the discharge boundary, called the *exit gradient*, which will yield the minimum resultant force (R_{min}) at this boundary.   This can be done analytically, as will be demonstrated later, or graphically from flow nets, after a method by Harza [54].   In the graphical method, the gradients along the discharge boundary are taken as the macrogradient across the contiguous squares of the flow net.   As the gradients along this boundary vary inversely with the distance between adjacent equipotential lines, it is evident that the maximum exit gradient is located where the vertical projection of this distance is a minimum, such as at the toe of the dam (point C) in Fig. 1-15.   For example, the head lost in the final square of Fig. 1-15 is one-sixteenth of the total head loss of 16 ft, or 1 ft, and, as this loss occurs in a vertical distance of approximately 4 ft, the exit gradient at point C is approximately 0.25.   Once the magnitude of the exit gradient has been found, the factor of safety with respect to piping is then ascertained by comparing this gradient with the critical gradient

* This is Risenkampf's *triangle of filtration* [122].

of Eqs. (8) or (9). For example, the factor of safety with respect to piping for the flow condition of Fig. 1-15 is 1.0/0.25 or 4.0. Factors of safety of 4 to 5 are generally considered reasonable for the graphical method of analysis.

1-14. Anisotropy

If the coefficient of permeability is independent of the direction of the velocity, the soil is said to be an *isotropic* flow medium. Moreover, if the soil has the same coefficient of permeability at all points within the region of flow, the soil is said to be *homogeneous* and *isotropic*. If the coefficient of permeability is dependent on the direction of the velocity and if this directional dependence is the same at all points of the flow region, the soil is said to be homogeneous and *anisotropic*. In homogeneous and anisotropic soils the coefficient of permeability is dependent on the direction of the velocity but independent of the space coordinates.

Most soils are anisotropic to some degree. Sedimentary soils often exhibit thin alternating layers. Stratification may result from particle orientation. Generally, in homogeneous natural deposits, the coefficient of permeability in the horizontal direction is greater than that in the vertical. One exception, worthy of special note, is loess, where, because of the vertical structure, the opposite is true.

Although Darcy's law was obtained initially from considerations of one-dimensional macroscopic flow only, in Sec. 1-9, upon the introduction of the velocity potential ϕ, it was demonstrated that the vectorial generalization of Darcy's law was valid for an isotropic flow medium. To provide a theoretical framework for any flow system it is necessary that this generalization take into account the directional dependence of the coefficient of permeability. Thus, it is generally assumed that

$$\mathbf{v}_n = -k_n \operatorname{grad}_n h \tag{1}$$

where k_n is the coefficient of permeability in the n direction and $\mathbf{v}_n$ and $\operatorname{grad}_n h$ are the components of the velocity and the hydraulic gradient in the same direction. For two-dimensional flow in the xy plane the velocity components in the x and y direction are

$$u = -k_x \operatorname{grad}_x h = -k_x \frac{\partial h}{\partial x}$$
$$v = -k_y \operatorname{grad}_y h = -k_y \frac{\partial h}{\partial y} \tag{2}$$

The work of this section will be divided into four parts: (1) It will be shown that a stratified medium of thin homogeneous and isotropic layers can be converted into an equivalent single homogeneous and isotropic layer. (2) It will be shown that the square root of the directional coefficient of permeability for an homogeneous and anisotropic layer when plotted from a point will generate an ellipse. (3) It will be shown that the effects of anisotropy can be taken into account by a simple transformation of spatial coordinates. (4) Finally, some aspects of nonhomogeneous systems will be considered.

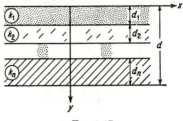

Fig. 1-17

1. Figure 1-17 represents a vertical section through a stratified soil of n thin isotropic layers of thickness $d_1, d_2, \ldots, d_n$, with coefficients of permeability, respectively, $k_1, k_2, \ldots, k_n$. For purely horizontal flow in the direction of stratification, the discharge through the sum of the layers will be

$$q = \sum_{m=1}^{n} k_m d_m \frac{h_1 - h_2}{L}$$

where $h_1 - h_2$ is the head lost by virtue of the flow through the layers in the distance L. The velocity in the x direction will be

$$u = \sum_{m=1}^{n} \frac{k_m d_m}{d} \frac{h_1 - h_2}{L}$$

and the equivalent coefficient of permeability in the x direction will be

$$k_x = \sum_{m=1}^{n} \frac{k_m d_m}{d} \tag{3}$$

From consideration the flow perpendicular to the direction of stratification, it is immediately apparent from the equation of continuity $\left(\dfrac{\partial v}{\partial y} = 0 \right)$ that the vertical velocity in all layers is the same. Thus,

$$v = k_y i = k_1 i_1 = k_2 i_2 = \cdots = k_n i_n \tag{4}$$

where i is the hydraulic gradient through the entire system of layers. Noting that the total head loss equals the sum of the head losses in each layer

$$id = i_1 d_1 + i_2 d_2 + \cdots + i_n d_n$$

we obtain

$$k_y = \frac{d}{\displaystyle\sum_{m=1}^{n} d_m / k_m} \tag{5}$$

It follows from Eqs. (3) and (5) that $k_x > k_y$; that is, the coefficient of permeability is greater in the direction of stratification. For simplicity, this will be verified for two layers, but it can be extended to any number of layers. Assuming $d_1/d_2 = \delta$, $k_x > k_y$ can be written as

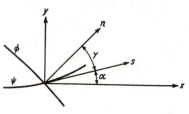

$$\frac{k_1\delta + k_2}{1 + \delta} > \frac{(\delta + 1)k_1k_2}{\delta k_2 + k_1} \qquad (6)$$

which reduces to the true statement

$$\delta(k_1 - k_2)^2 > 0 \qquad (7)$$

FIG. 1-18

2. In Fig. 1-18, let s and n represent the directions of the tangent to the flow line and the normal to the equipotential line, respectively. In an isotropic flow medium the flow lines and equipotential lines form an orthogonal system; hence the s and n directions are identical. In anisotropic flow, however, as will be shown below, the direction of the streamlines will not, in general, coincide with the direction of the normal to the equipotential lines.

The resultant velocity along the streamline in Fig. 1-18 is

$$v_s = -k_s \frac{\partial h}{\partial s} \qquad (8a)$$

and hence the velocity components in the x and y directions are

$$u = -k_x \frac{\partial h}{\partial x} = v_s \cos \alpha \qquad (8b)$$

$$v = -k_y \frac{\partial h}{\partial y} = v_s \sin \alpha \qquad (8c)$$

Since

$$\frac{\partial h}{\partial s} = \frac{\partial h}{\partial x} \frac{\partial x}{\partial s} + \frac{\partial h}{\partial y} \frac{\partial y}{\partial s}$$

we find, substituting Eqs. (8), that

$$\frac{1}{k_s} = \frac{\cos^2 \alpha}{k_x} + \frac{\sin^2 \alpha}{k_y} \qquad (9a)$$

or

$$k_s = \frac{k_x k_y}{k_x \sin^2 \alpha + k_y \cos^2 \alpha} \qquad (9b)$$

Converting Eq. (9a) into rectangular coordinates ($x = r \cos \alpha$, $y = r \sin \alpha$) we have

$$\frac{r^2}{k_s} = \frac{x^2}{k_x} + \frac{y^2}{k_y} \qquad (10)$$

which is the equation of an ellipse with major and minor semiaxes of $k_x^{1/2}$ and $k_y^{1/2}$. If $k_x = k_{max}$ and $k_y = k_{min}$, as in the case of a stratified medium with the x direction parallel to the bedding plane, the coefficient of permeability in any direction (making an angle α with the x axis) can be obtained easily from the graphical construction of Fig. 1-19, called *the ellipse of direction.*

Although no formal proof of the generalization of Darcy's law in an anisotropic flow medium exists, Scheidegger's results [127] of physical measurements of directional permeability tend to substantiate the validity of Eqs. (9) and to verify the existence of the ellipse of direction.

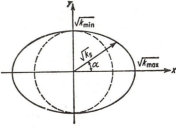

3. Equation (10) and Fig. 1-19 demonstrate that by the transformation of scale in the y direction of $Y = y(k_x/k_y)^{1/2}$, the ellipse of direction will

Fig. 1-19

be transformed into a circle wherein the coefficient of permeability will be an invariant with direction. A similar transformation can be achieved in the x direction with $X = x(k_y/k_x)^{1/2}$ (dotted circle in Fig. 1-19). This in essence provides the method for solving problems in homogeneous and anisotropic layers. The validity of the transformation of coordinates can be demonstrated in a more rigorous manner directly from the equation of continuity

$$\frac{\partial u}{\partial x} + \frac{\partial v}{\partial y} = 0$$

Substituting for u and v from Eqs. (8b) and (8c), we have

$$\frac{\partial^2 h}{\partial[(k_y/k_x)x^2]} + \frac{\partial^2 h}{\partial y^2} = 0$$

which by transformation of coordinates [say $X = x(k_y/k_x)^{1/2}$] yields at once

$$\frac{\partial^2 h}{\partial X^2} + \frac{\partial^2 h}{\partial y^2} = 0 \qquad (11)$$

Obviously, a similar transformation can be effected in the y direction.

The results of this subsection indicate that by a simple expansion or contraction of spatial coordinates a given homogeneous and anisotropic flow region can be transformed into a fictitious isotropic region wherein Laplace's equation is valid and consequently the potential theory is applicable. This fictitious flow region is called the *transformed section.* Once the problem has been solved for the transformed section, the solution for the natural flow medium can be obtained by applying the inverse of the scaling ratio. An example of the transformed section is shown by the dotted lines in Fig. 1-20. The solid lines define the natural section, which in this case represents a cross section through a row of impervious

sheet piles founded in an anisotropic base. The permeability characteristics of the base soil is provided by an ellipse of permeability inclined
$45°$ to the horizontal with $k_1 = 4k_2$. To obtain the transformed section
a coordinate system, parallel to the main axes of the ellipse, is established
through some point in the flow domain (point 2 in Fig. 1-20a). The

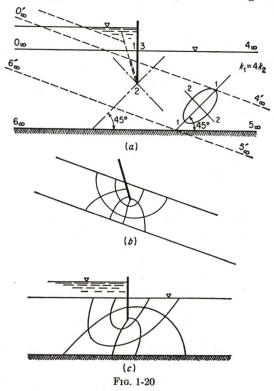

Fig. 1-20

boundaries of the fictitious flow region, wherein Laplace's equation is
valid, are then obtained by multiplying the perpendicular distances from
the boundaries of the natural flow domain to the reference axes by the
approximate scaling factor. In Fig. 1-20a all distances parallel to the
1-1 axis were reduced by $(k_2/k_1)^{1/2}$. The primed numbers locate the
vertices of the transformed section. Once the flow problem has been
solved for the transformed section (the transformed flow net is given in
Fig. 1-20b), by applying the inverse of the scaling factor $[(k_1/k_2)^{1/2} = 2]$
the solution is obtained for the natural boundaries (Fig. 1-20c).

The equivalent coefficient of permeability for an homogeneous and
anisotropic section is

$$k = \sqrt{k_{max}k_{min}} \qquad (12)$$

A formal derivation of Eq. (12) can be found in Refs. 88, 159; a shorter but less rigorous proof follows. Consider in Fig. 1-21a the transformed section, a curvilinear square bounded by the equipotential lines ϕ_1 and ϕ_2 and the streamlines ψ_1 and ψ_2; $k_x = k_{max}$, $k_y = k_{min}$. Figure 1-21b represents the same section to the natural scale. Vertical line AB, parallel to the y direction in the transformed section, becomes

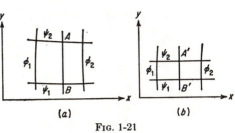

$$A'B' = AB\left(\frac{k_{min}}{k_{max}}\right)^{\frac{1}{2}}$$

(a)   (b)

Fig. 1-21

to the natural scale. The quantity of flow across section AB must be equal to that across $A'B'$ ($\Delta\psi$ = constant). Hence, in the transformed section,

$$q = ABk \text{ grad } h$$

and in the natural section, since the coefficient of permeability in the x direction is k_{max},

$$q = A'B'k_{max} \text{ grad } h = k_{max}\sqrt{\frac{k_{min}}{k_{max}}} AB \text{ grad } h$$

Equating the two expressions for q, we obtain Eq. (12).

Let us consider the nature of a flow line at the boundary AB between two isotropic soil of permeabilities k_1 and k_2 (Fig. 1-22). Designating the velocity potentials within the respective zones as

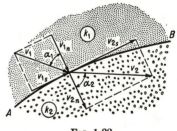

$$\phi_1 = -k_1\left(\frac{p_1}{\gamma_w} + y_1\right)$$

$$\phi_2 = -k_2\left(\frac{p_2}{\gamma_w} + y_2\right)$$

Fig. 1-22

we see that at all points along the interface ($p_1 = p_2$, $y_1 = y_2$)

$$\frac{\phi_1}{k_1} = \frac{\phi_2}{k_2} \tag{13}$$

From the equation of continuity, the normal components of the vectors along the boundary AB must be equal; hence

$$v_{1_n} = v_{2_n} \tag{14}$$

Now, differentiating Eq. (13) with respect to the length of arc s and dividing by Eq. (14), we find

$$\frac{v_{1_s}}{v_{1_n}k_1} = \frac{v_{2_s}}{v_{2_n}k_2}$$

which, since $v_s/v_n = \tan \alpha$, yields

$$\frac{k_1}{k_2} = \frac{\tan \alpha_1}{\tan \alpha_2} \tag{15}$$

The similarity between Eq. (15) and the *law of incidence and refraction* in optics is obvious.

From Eq. (13) we note that the relative spacing of the equipotential lines in the two zones is dependent on the ratio of their respective permeabilities. For example, if $k_2 = 3k_1$, a flow net drawn with curvilinear squares in zone 1 will exhibit curvilinear rectangles in zone 2 with the distance between adjacent equipotential lines being 3 times ($k_2/k_1 = 3$) the width of the flow channel. A portion of a flow net with $k_2 = 3k_1$ is shown in Fig. 1-23.

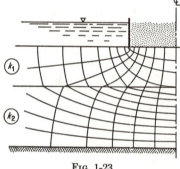

FIG. 1-23

If we consider each of the soils in Fig. 1-23 to be homogeneous and anisotropic, two possibilities, depending upon the orientation of their ellipses of permeability, must be investigated. In the first case it will be assumed that the orientations of the ellipses in each of the zones are the same, although the ratio of $k_{\max}/k_{\min}$ in each may differ. Recalling that an homogeneous and anisotropic medium can be transformed into an equivalent fictitious isotropic one, we may transform the boundaries of each of the regions in a direction normal to the interface and so obtain a transformed section with two homogeneous and isotropic zones, as was considered above.* Although no rigorous solution exists for the second possibility, flow regions with two homogeneous and anisotropic zones with different orientations of the ellipses of permeability, an approximate method proposed by Stevens [140] which transforms the two zones into equivalent homogeneous and isotropic zones is worthy of consideration. In Fig. 1-24a, the anisotropy of zone I is defined by the ellipse of direction O_1 with the semiaxes $k_1^{1/2}$ and $k_2^{1/2}$ and in zone II by the ellipse of direction O_2 with the semiaxes $(k_1')^{1/2}$ and $(k_2')^{1/2}$. The procedure is as follows: The

---

* A simple but elegant analytical treatment of these problems will be given in Sec. 6-7.

ellipse of direction in zone I is transformed into a circle and the boundaries of the region are expanded by applying the scaling factor $(k_1/k_2)^{1/2}$ in the 2-2 direction, holding point G as the reference point.   The new zone I is shown dashed in Fig. 1-24b and the new vertices of the region are shown

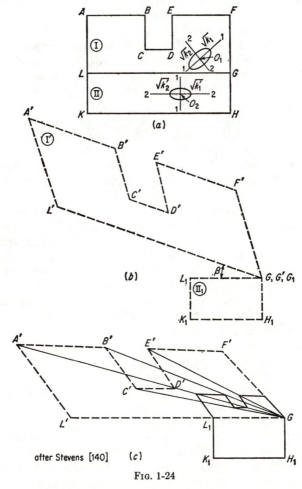

after Stevens [140]   (c)

Fig. 1-24

primed.   Zone II is transformed in the 2-2 direction, again keeping point G as the reference point, so that the figure $G_1H_1K_1L_1$ is obtained.   As a result of these transformations a discontinuity is developed at the interface of the layers.   To rectify this, Stevens recommends that zone I be rotated through the angle β about point G, so that the lines $L'G'$ and L_1G_1 coincide.   At the same time all radial distances from point G in zone I' are multiplied by a factor $\mu = GL_1/GL'$ so that L' coincides with L_1. The resulting figure is shown solid in Fig. 1-24c. Thus the regions are once again reduced to an equivalent homogeneous and isotropic system, with known boundary conditions, wherein potential theory is applicable. The same procedure can be extended to any number of zones. Once the problem is solved in the transformed section, the true nature of the flow can be obtained by reversing the procedure.

The foregoing work demonstrates that, in theory, any homogeneous flow system, whether sectionally isotropic or not, can be transformed into a tractable flow medium. Also, if a nonhomogeneous soil can be considered as an aggregate of thin, alternating homogeneous and isotropic layers with varying permeabilities (even varying thicknesses), the soil can be transformed into an equivalent homogeneous and anisotropic soil. The transformation from an homogeneous and anisotropic soil into a fictitious isotropic soil was demonstrated with some rigor.

In the construction of fills most soils are placed and compacted in horizontal lifts, and hence are likely to fall into the category of homogeneous and anisotropic layered systems wherein $k_{horizontal} > k_{vertical}$. In the construction of earth structures to retain water, such stratification is undesirable and special efforts are generally taken during construction to reduce it. Sheepsfoot rollers are generally effective in minimizing stratification. However, Casagrande [14] notes that even the most carefully constructed rolled-earth dams possess a considerably greater average permeability in a horizontal than in a vertical direction.

4. In some natural soil deposits the degree of heterogeneity is so great that any transformation of the individual zonal permeabilities is impractical. In such cases it is generally expedient to neglect the localized variations in favor of a study of the overall characteristics of the medium. Many simplifications may be necessary to gain insight into the pertinent flow characteristics of the foundation. Several schemes may be investigated wherein the influence of combinations of some of the zonal deposits are considered and others are neglected. Frequently, and as a first approximation, overall typical values may be assumed for horizontal and vertical permeabilities, and the section may be treated as an effective homogeneous and anisotropic layer or layers.

To illustrate the avenues of approach when dealing with complicated foundations, consider the cross section of an earth dam and its base given in Fig. 1-25. Depending upon the ratio of the coefficients of permeability of the various regions, an effective impervious boundary can be located at, say, either of the solid lines **1-1** or **2-2**. In this regard, when the ratio of the permeabilities between neighboring soils is greater than about 10 to 1, the soil of lesser permeability may be taken as impervious. Thus, if $k_5/k_4 > 10$, **1-1** can be considered as impervious; or, if $k_4/k_7 > 10$,

2-2 may be taken as such. If the quantity of seepage through the base soils of Fig. 1-25 is desired (assuming k_1 impervious), the section might be

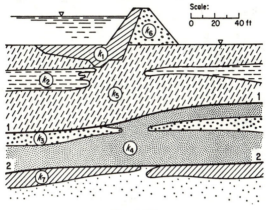

FIG. 1-25

investigated by considering various combinations of equivalent homogeneous isotropic layers. For example, **1-1** may be assumed as an impervious lower boundary with the soil above it characterized by a coefficient of permeability of k_5 (neglecting seams of k_2); or, a two-layered system of k_5 and k_4 soils with **2-2** as an impervious boundary (neglecting both the k_2 and k_3 seams) may be investigated. One might also consider the possibility of an equivalent single anisotropic layer with an horizontal coefficient of permeability equal to some combination of k_2 and k_3 and a vertical permeability dependent upon k_4 and k_5 with a depth of layer intermediate between **1-1** and **2-2**.

The work of this section demonstrates that any two-dimensional flow system, regardless of its degree of homogeneity and isotropy, can be converted into an equivalent flow domain wherein the salient aspects of seepage are tractable. Thus we need concern ourselves only with homogeneous and isotropic systems; in our subsequent work unless otherwise stated it will be assumed that the cross sections under consideration have been transformed a priori and that the related coefficients of permeability aptly typify the porous nature of the considered regions.

PROBLEMS

1. (a) How many spheres touch the shaded sphere in Fig. 1-1b? (b) Estimate the porosity of the sphere packing if $\beta = 75°$.

2. A cylinder of soil 6 in. in height exhibits an effective ratio of the area of pores that varies as $\cos(\pi z/3h)$, where z and h are as given in Fig. 1-4. (a) Compute the volume porosity of the soil sample. (b) Compute the volume porosity if the cylinder is 12 in. in height.

3. Demonstrate that the pressure at any point within an ideal fluid (incompressible and nonviscous) is an invariant with direction even if the fluid is accelerating.

4. Derive the forms of the following by the Buckingham pi theorem [97]:

(a) Darcy's law

(b) Reynolds number

(c) Equation 3, Sec. 1-6

5. Describe a common engineering mechanism where, in Bernoulli's equation, in comparison to other factors, (a) the pressure head is negligible; (b) the elevation head is negligible.

6. Demonstrate that Reynolds number is the ratio of inertial force to viscous force.

7. If the discharge velocity in a constant-head permeameter test with water is 0.25 cm/sec at 80°F, what would it be at 130°F? At 5°F?

8. If water rises in x minutes to 8 per cent of the maximum height of capillary rise in a circular tube filled with uniform soil, how long will it take to rise the same height in the same tube if the soil is of twice the diameter? Assume the contact angle is zero and that k varies directly with the square of the diameter of the soil particles.

9. Verify Eq. (2c), Sec. 1-8.

10. Introducing the gravity potential $\Omega = gz$ into Eqs. (2), Sec. 1-8, and defining the gravity force potentials in the x, y, and z directions as

$$X = -\frac{\partial \Omega}{\partial x} \qquad Y = -\frac{\partial \Omega}{\partial y} \qquad Z - g = -\frac{\partial \Omega}{\partial z}$$

demonstrate that the integrated form of Euler's equations yields the Bernoulli equation for steady-state flow.

11. Derive the equation of continuity in polar form for two-dimensional flow:

$$\frac{\partial(rv_r)}{\partial r} + \frac{\partial v_\theta}{\partial \theta} = 0$$

If $v_\theta = -(c \sin \theta)/r^2$, where c is a constant, determine the value of v_r and the magnitude of the resultant velocity.

12. If $\psi = cx^n$, where c and n are integers, determine the values of n which would assure the flow to be irrotational. Continuous. Find the velocity components at point $(1,1)$.

13. Determine whether the following stream functions guarantee irrotational flow:

(a) $\psi = \sinh x$ (b) $\psi = \sin^{-1} \frac{x}{y}$ (c) $\psi = 2cxy$

(d) $\psi = \frac{c}{\text{mod } z}$ (e) $\psi = \ln x$

14. The velocity potential for a particular flow condition is $\phi = x^2 - y^2 = $ constant. Determine the corresponding stream function and the magnitude and direction of the resultant velocity at the points $(1,0)$, $(1,1)$, and $(2,0)$.

15. Demonstrate that if $w = \phi + i\psi$ and $z = x + iy, f'(z) = u - iv, f'(z) = dw/dz$. Discuss the nature of the functions $|f'(z)|$, $f'(z)$, and arg $f'(z)$.

16. (a) Demonstrate that along the boundaries of a reservoir (Fig. 1-11b) the pressure distribution is hydrostatic. (b) Demonstrate that the surface of seepage can be neither an equipotential line nor a streamline.

17. Demonstrate that Eq. (4), Sec. 1-12, can be obtained directly from the statement of Darcy's law.

18. If $k = 1 \times 10^{-2}$ cm/sec, obtain the magnitude and direction of the velocity at points 1, 2, and 3 in Fig. 1-15.

19. Determine the required length of impervious upstream blanket x in Fig. 1-26 to reduce the quantity of seepage by 35 per cent.

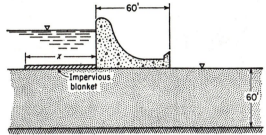

FIG. 1-26

20. By constructing flow nets for the section of Fig. 1-27, obtain an empirical expression for the reduced quantity of seepage as a function of the ratio s/T and h.

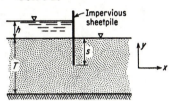

FIG. 1-27

21. Repeat the procedure of Prob. 20 to obtain an expression for the exit gradient i_E as a function of the ratio s/T in the form $i_E s/h = f(s/T)$.

22. Estimate the magnitude of the uplift force acting on the base of the structure in Fig. 1-15.

23. Obtain the plot showing the distribution of the factor of safety with respect to piping along CD of Fig. 1-15. Employing a soil with $\gamma_m = 1.2\gamma_{m_1} = 2\gamma_w$ as a filter, where γ_{m_1} is the mass unit weight of the base soil in Fig. 1-15, design a reverse filter such that the factor of safety with respect to piping along CD is everywhere greater than 7.

24. For a particular pattern of groundwater flow the streamlines are confocal ellipses symmetrical about the origin. The x axis denotes the ground surface. At $x < 0$ the pressure on the ground surface is 624 psf; at $x > 0$, 62.4 psf. Estimate the direction and magnitude of the resultant force acting at the points $x = -2$, 0, +4, and +6 along the streamline $x^2 + 4y^2 = 36$. γ_m of soil $= 2\gamma_w$; $S_s = 2.65$.

25. Three thin horizontal homogeneous isotropic layers of equal thickness are subject to laminar flow. The coefficient of permeability of the top layer is $4k$, of the middle layer $2k$, and of the bottom layer k. If the layered system is converted into one equivalent homogeneous anisotropic layer, determine the theoretical coefficient of permeability at an angle of 30° from the horizontal. What scaling factor applied to the horizontal direction would convert the layer into an equivalent isotropic layer?

26. (a) Demonstrate, in any homogeneous flow medium (isotropic or anisotropic), that wherever the tangent to a streamline is in the direction of k_{max} or k_{min}, the equipotential line passing through that point intersects the streamline orthogonally. (b) What is the direction of the seepage force at any point in anisotropic flow? (c) Is it possible to determine the quantity of seepage and the factor of safety with respect to piping directly from a transformed section?

27. Is it possible in an homogeneous anisotropic layer to determine the coefficient of permeability in any direction if it is known in any two directions?

28. Determine the required expressions in Probs. 20 and 21 assuming the base soil in Fig. 1-27 to be homogeneous with $k_x = k_{max}$ and $k_y = k_{min}$.

29. Obtain the flow nets for each of the sections in Fig. 1-28 for both the natural and transformed sections.

30. Determine the quantity of seepage and the factor of safety with respect to piping for each of the flow conditions of Fig. 1-28.

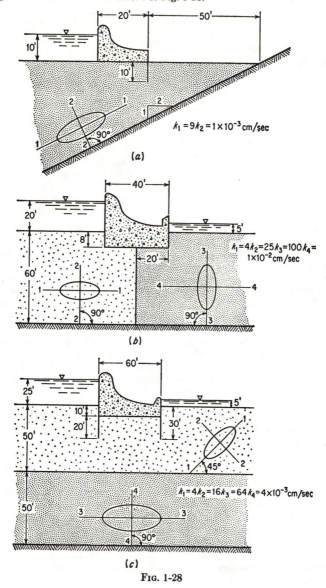

Fig. 1-28

31. Find the general relationship among α, θ, k_1, and k_2 that would make the inclined sheetpile in Fig. 1-29 in the transformed section perpendicular to the ground surface.

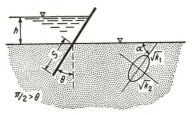

FIG. 1-29

32. In Fig. 1-25, $k_1 = k_7 = 1 \times 10^{-6}$ cm/sec, $k_2 = 1 \times 10^{-3}$ cm/sec, $k_3 = 1 \times 10^{-4}$ cm/sec, $k_4 = 1 \times 10^{-5}$ cm/sec, $k_5 = 1 \times 10^{-1}$ cm/sec, and $k_6 = 1$ cm/sec. Estimate the maximum quantity of seepage through the section.

33. Repeat Prob. 32 with $k_4 = k_5 = 1 \times 10^{-1}$ cm/sec.

34. Repeat Prob. 32 with $k_2 = k_6 = 1 \times 10^{-3}$ cm/sec.

2

Application of the Dupuit Theory of Unconfined Flow

2-1. Basic Considerations

The class of problems to be considered in this chapter is characterized by having one boundary of the flow domain a free surface (Sec. 1-11). Such flow patterns (or domains) are said to be *unconfined*. This is in contrast to those flow domains where all the boundaries are known initially and the flow is said to be *confined*.

The *Dupuit theory* of unconfined flow stems from two assumptions first made by Dupuit [28] in 1863. Dupuit assumed: (1) that for small inclinations of the line of seepage the streamlines can be taken as horizontal (hence the equipotential lines approach the vertical), and (2) that the hydraulic gradient was equal to the slope of the free surface and was invariant with depth. Although the nature of these assumptions appear paradoxical, in many groundwater problems solutions based on the Dupuit assumptions compare favorably with those of more rigorous methods.

From Eq. (11), Sec. 1-8, we have for the relationship between the velocity potential ϕ and the total head h,

$$\phi(x,y,z) = -kh(x,y,z) = -k\left(z + \frac{p}{\gamma_w}\right)$$

Considering atmospheric pressure as zero, we obtain at the free surface, with coordinate axes as in Fig. 2-1,

$$h(x,y,z) = z \tag{1}$$

or
$$\phi(x,y,z) + kz = 0$$

Assuming that the free surface will vary but little from some average value $\bar{z}$ and expanding $h(x,y,z)$ in Taylor's series about $\bar{z}$, we obtain

$$h(x,y,z) = h(x,y,\bar{z}) + \left(\frac{\partial h}{\partial z}\right)_{z=\bar{z}} (z - \bar{z}) + \cdots \tag{2}$$

Now, if the vertical velocity is not exceedingly large (that is, if $\partial h/\partial z$, which is proportional to the vertical velocity, is small) and the free surface is relatively level [$(z - \bar{z})$ is of small order], Eq. (2) can be taken as $h(x,y,z) = h(x,y,\bar{z})$. As $\bar{z}$ is a constant equal to the average depth of flow but independent of the z coordinate, the total head becomes a function of the variables x and y only, or

$$h = h(x,y) \tag{3}$$

Equation (3) demonstrates that the lines of equal total head, and hence the equipotential lines, are vertical, as stated in Dupuit's assumption 1. Now, substituting Eq. (3) for the left side of Eq. (1)

$$h(x,y) = z \tag{4}$$

we see that the total head along any vertical line is a constant equal to the elevation of the free surface at that line. Hence the hydraulic gradient along any vertical line is equal to the slope of the free surface and is an invariant with depth, as specified in assumption 2. Thus, the validity of Dupuit's assumptions in a given flow situation is highly contingent on the steepness of the line of seepage. In this chapter we shall investigate various flow domains using Dupuit's simplifying assumptions. In a later chapter problems of unconfined flow will be treated in a more rigorous manner.

Let Fig. 2-1 represent a vertical column of fluid, bounded from above by the free surface and from below by the horizontal xy plane, which will be assumed to be impervious. Making use of the horizontal components of the velocity

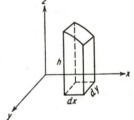

FIG. 2-1

$$u = -k \frac{\partial h}{\partial x} \qquad v = -k \frac{\partial h}{\partial y}$$

we see that the discharge q_x in the x direction (per unit width in the y direction) through the left face of the column is the product of the area $h \, dy$ and the velocity u or

$$q_x \, dy = -k \left(h \frac{\partial h}{\partial x} \right)_x dy \tag{5}$$

The discharge through the right face q_{x+dx} is

$$q_{x+dx} \, dy = -k \left(h \frac{\partial h}{\partial x} \right)_{x+dx} dy \tag{6}$$

and hence the change in the quantity of flow in the x direction is

$$(q_{x+dx} - q_x) \, dy = \frac{\partial q_x}{\partial x} dx \, dy = -k \frac{\partial}{\partial x} \left(h \frac{\partial h}{\partial x} \right) dx \, dy \tag{7}$$

By a similar procedure, the change in the quantity of flow in the y direction is found to be

$$\frac{\partial q_y}{\partial y}\, dx\, dy = -k\frac{\partial}{\partial y}\left(h\frac{\partial h}{\partial y}\right) dx\, dy \tag{8}$$

If it is assumed that we have saturated, incompressible steady-state flow, the sum of the changes in the quantity of flow must be identically zero; hence

$$k\left[\frac{\partial}{\partial x}\left(h\frac{\partial h}{\partial x}\right) + \frac{\partial}{\partial y}\left(h\frac{\partial h}{\partial y}\right)\right] = 0 \tag{9}$$

and

$$\frac{\partial^2(h^2)}{\partial x^2} + \frac{\partial^2(h^2)}{\partial y^2} = 0 \tag{10}$$

Therefore, for the Dupuit theory, the function h^2 must satisfy Laplace's equation. This result was first demonstrated by Forchheimer [37], although the presentation given here follows more closely that of Polubarinova-Kochina [116].

2-2. Two-dimensional Flow on a Horizontal Impervious Boundary

Consider the xz plane of Fig. 2-2 as the plane of flow. h_1 and h_2 represent the elevations of two known points on the free surface. For this

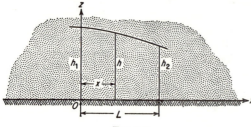

Fig. 2-2

case Eq. (10), Sec. 2-1 reduces to

$$\frac{d^2(h^2)}{dx^2} = 0 \tag{1}$$

which, after integration, gives the parabola

$$h^2 = Ax + B \tag{2}$$

where A and B are constants. Now, applying the boundary conditions $x = 0$, $h = h_1$, and $x = L$, $h = h_2$, we obtain the expression for the elevation h at any intermediate point.

$$h = \sqrt{h_1{}^2 - (h_1{}^2 - h_2{}^2)\frac{x}{L}} \tag{3}$$

The discharge (per unit width) through any vertical section in Fig. 2-2 is

$$q = -kh\frac{dh}{dx} \tag{4}$$

which, after integration and substitution of the boundary conditions, yields

$$q = k\frac{h_1{}^2 - h_2{}^2}{2L} \tag{5}$$

Equation (5) is called *Dupuit's formula.*

Let us investigate the nature of the vertical velocities which we previously assumed to be negligible. Writing the equation of continuity in two dimensions as

$$\frac{\partial u}{\partial x} + \frac{\partial w}{\partial z} = 0$$

and integrating with respect to z (recalling that u is a function of x only, $u = -k\, dh/dx$), we obtain

$$w = -\int_0^z \frac{\partial u}{\partial x}\, dz = k \int_0^z \frac{d^2 h}{dx^2}\, dz$$

As $d^2 h/dx^2$ is not a function of z, and at the impervious boundary w must equal zero,

$$w = kz \frac{d^2 h}{dx^2} \tag{6}$$

which, with the use of Eq. (4), reduces to

$$w = -\frac{q^2 z}{kh^3} \tag{7}$$

At the free surface, where $z = h$, the vertical velocity is

$$w = -\frac{q^2}{kh^2} = -k\left(\frac{q}{kh}\right)^2 \tag{8}$$

Equations (7) and (8) demonstrate that the vertical velocity is down (negative) and increases with elevation.

2-3. Free Surface Subject to Infiltration or Evaporation [66]

Assume that the free surface in Fig. 2-3 is subject to a uniform discharge e per unit area ($e > 0$ in the case of infiltration, $e < 0$ for evaporation).

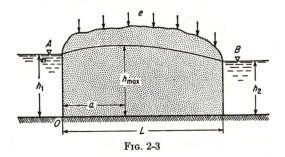

FIG. 2-3

With flow in the xz plane, Eq. (9), Sec. 2-1, becomes

$$k \frac{d}{dx}\left(h \frac{dh}{dx}\right) + e = 0 \tag{1}$$

After integration, this takes the form

$$kh^2 + ex^2 = C_1x + C_2 \tag{2}$$

For infiltration $(e > 0)$ Eq. (2) is an ellipse, whereas in the case of evaporation $[e = -\varepsilon(\varepsilon > 0)]$ it is the hyperbola

$$kh^2 - \varepsilon x^2 = C_1x + C_2 \tag{3}$$

Considering infiltration, we can obtain several useful approximate relationships. Substituting the boundary conditions $x = 0$, $h = h_1$ and $x = L$, $h = h_2$ into Eq. (2), we find for the equation of the free surface

$$h = \sqrt{h_1^2 - \frac{(h_1^2 - h_2^2)x}{L} + \frac{e}{k}(L - x)x} \tag{4}$$

Now, to determine the discharge in the case of infiltration, Eq. (1) is written as

$$\frac{dq_x}{dx} = e$$

which, after integration, yields

$$q_x = ex + q_1 \tag{5}$$

where q_x is the quantity of seepage at any vertical section, and q_1 is the quantity of seepage at $x = 0$. Substituting for q_x from Eq. (4), Sec. 2-2, and integrating with the boundary conditions of Eq. (4) of this section, we obtain

$$q_1 = \frac{k(h_1^2 - h_2^2)}{2L} - \frac{eL}{2}$$

which, when substituted back into Eq. (5), gives finally

$$q_x = \frac{k(h_1^2 - h_2^2)}{2L} - e\left(\frac{L}{2} - x\right) \tag{6}$$

Designating the distance to the maximum elevation of the free surface in Fig. 2-3 as a, called the *water divide*, we find, from Eq. (6) with $q_x = 0$,

$$a = \frac{L}{2} - \frac{k}{e}\frac{h_1^2 - h_2^2}{2L} \tag{7}$$

2-4. Groundwater Flow with an Inclined Lower Impervious Boundary

Although the problems to be considered in Secs. 2-4 to 2-6 may appear similar to those of open-channel flow [19], a very basic difference between the two should be noted. Whereas most open-channel flows are turbulent and hence require complicated analyses, with groundwater flow, because of the small velocities involved, the flow is laminar and the resulting solutions are greatly simplified. The methods of analysis given here were first published by Pavlovsky [110] in 1930.

Consider the impervious boundary in Fig. 2-4 to be inclined at some small angle α with the horizontal axis. Designating the ordinate of the free surface as y (distance from the horizontal) and defining $i = \tan \alpha$, the quantity of seepage through any section (such as AB) becomes

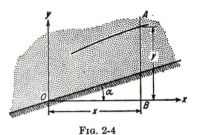

$$q = -k(y - ix)\frac{dy}{dx} \qquad (1)$$

Rewriting Eq. (1) as

$$\frac{dx}{dy} - \frac{kix}{q} + \frac{k}{q}y = 0$$

Fig. 2-4

and noting that continuity requires the quantity of seepage to be constant, we find

$$y - ix + \frac{q}{ki} = C \exp \frac{kiy}{q} \qquad (2)$$

If $C = 0$ in Eq. (2), we have the particular solution

$$y - ix = -\frac{q}{ki} = \text{constant} \qquad (3)$$

which corresponds to seepage with a constant velocity parallel to the impervious boundary (line I of Fig. 2-5). The flow in this case is called *uniform flow* and the constant depth of the stream $h_0 = y - ix$ is called the *normal depth*. The quantity of seepage for uniform flow is simply

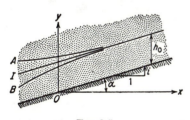

Fig. 2-5

$$q = -kh_0 i \qquad (4)$$

For values of $C \neq 0$ the flow is called *nonuniform*. Equation (2) for this case is seen to be asymptotic to Eq. (3), and hence the quantity of seepage for both uniform and nonuniform flow will be given by Eq. (4). Thus Eq. (2) can be rewritten as

$$y - ix - h_0 = C \exp\left(\frac{-y}{h_0}\right) \qquad (5)$$

Three possible conditions need to be investigated to determine the shape

of the free surface. As shown in Fig. 2-6, these depend on the orientation of the impervious boundary and the relationship between headwater and tail-water elevations.

If $i > 0$ and $C > 0$, it is evident from Eq. (5) that the free surface will lie entirely above the normal depth (that is, $y - ix > h_0$) with asymptote at $y = \infty$ (curve A of Fig. 2-5). For this case, corresponding to Fig. 2-6a, the free surface is said to be a *rising surface*. If $i > 0$ and $C < 0$, from Eq. (5) we see that the free surface will be completely below the normal depth (curve B of Fig. 2-5); this corresponds to the condition shown in Fig. 2-6b and is said to be a *falling surface*. Now if $i < 0$ (Fig. 2-7), Eq. (4) becomes $q = kih_0$ and Eq. (5) becomes

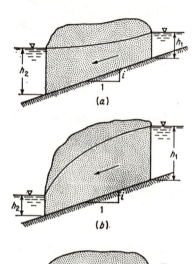

$$y + ix + h_0 = C \exp\left(\frac{y}{h_0}\right)$$

Thus we see that the free surface again tends to the asymptote

$$y = -ix - h_0$$

However, in this case the asymptote exists below the impervious bound-

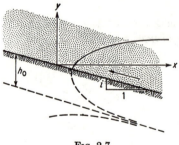

(a)

(b)

(c)

Fig. 2-6   Fig. 2-7

ary (for negative values of y). As only the falling free surface (solid curve in Fig. 2-7) can exist above the impervious boundary and have any physical significance, it represents the only solution for the flow situation shown in Fig. 2-6c.

If a point on the free surface (x_1, y_1) is known, the constant C in Eq. (5) can be found from

$$y_1 = ix_1 - h_0 = C \exp\left(-\frac{y_1}{h_0}\right) \qquad (6)$$

Excluding C from Eqs. (2) and (6), we get

$$y - ix - h_0 = (y_1 - ix_1 - h_0) \exp\left(-\frac{y - y_1}{h_0}\right) \tag{7}$$

Now if an additional point on the free surface (x_2, y_2) is known, Eq. (7) will reduce to

$$y_2 - y_1 = h_0 \ln \frac{y_1 - ix_1 - h_0}{y_2 - ix_2 - h_0} \tag{8}$$

from which h_0 and hence the free surface and the quantity of seepage can be found. The work necessary to effect a solution can be greatly reduced by making a change of variables.

2-5. Pavlovsky's Solution for $i > 0$

To simplify the solution of the previous section, Pavlovsky [110] introduced the new variables

$$h = y - ix \tag{1}$$

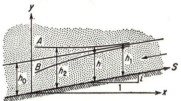

Fig. 2-8

and S, as shown in Fig. 2-8. For small values of i, S can be taken as equal to x. Thus, substituting Eq. (1) into Eqs. (7) and (8), Sec. 2-4, and assuming i to be small as specified in the Dupuit assumptions, we get, respectively,

$$h - h_1 + i(x - x_1) = h_0 \ln \frac{h_1 - h_0}{h - h_0} \tag{2}$$

$$h_2 - h_1 + i(x_2 - x_1) = h_0 \ln \frac{h_1 - h_0}{h_2 - h_0} \tag{3}$$

Now defining $\eta = h/h_0$, $\eta_1 = h_1/h_0$, $\eta_2 = h_2/h_0$, and $L = x_1 - x_2$, Eq. (3) reduces to

$$\frac{iL}{h_0} = \eta_2 - \eta_1 + \ln \frac{\eta_2 - 1}{\eta_1 - 1} \qquad \text{for } \eta > 1 \tag{4}$$

and

$$\frac{iL}{h_0} = \eta_2 - \eta_1 + \ln \frac{1 - \eta_2}{1 - \eta_1} \qquad \text{for } \eta < 1 \tag{5}$$

Equation (4) is applicable to rising free surfaces (Fig. 2-6a), and Eq. (5) to falling free surfaces (Fig. 2-6b).

Now if we define

$$\begin{aligned}
\phi(\eta) &= \eta + \ln (\eta - 1) &\qquad \text{for } \eta > 1 \\
\phi(\eta) &= \eta + \ln (1 - \eta) &\qquad \text{for } \eta < 1
\end{aligned} \tag{6}$$

then Eqs. (4) and (5) both take the form

$$\frac{iL}{h_0} = \phi(\eta_2) - \phi(\eta_1) \tag{7}$$

which greatly simplifies the solution.   The $\phi(\eta)$ functions are plotted in Fig. 2-9 as a function of η.   Should more accuracy be required they can be obtained directly from Eqs. (6).

The method of solution will now be illustrated by an example.

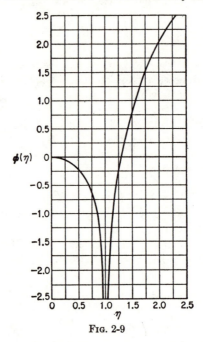

Fig. 2-9

Example 2-1.   A canal 2 km long flows parallel to a river 300 meters away.   The depth of water in the canal is $h_1 = 2$ meters, the depth of the river is $h_2 = 4$ meters. The impervious boundary is inclined $i = +0.025$ and the coefficient of permeability of the soil is $k = 0.002$ cm/sec (Fig. 2-10).   Construct the line of seepage and determine the quantity of seepage from the canal.

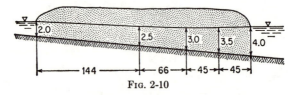

Fig. 2-10

Since $i > 0$ and $h_2 > h_1$, we know that the free surface will be of the rising type. Substituting given values into Eq. (7), we have

$$\frac{0.025}{h_0} 300 = \phi(\eta_2) - \phi(\eta_1)$$

or

$$7.5 = h_0[\phi(\eta_2) - \phi(\eta_1)] = f(h_0)$$

Assuming values of h_0 of 1.85, 1.9, and 1.95 meters [as h_0 was shown to be below the free surface for the rising type (Sec. 2-4)] and obtaining $\phi(h/h_0)$ values from Fig. 2-9 or Eq. (6), we find the corresponding values of $f(h_0)$ given in Table 2-1.

Table 2-1

h_0	$\dfrac{h_2}{h_0}$	$\phi\left(\dfrac{h_2}{h_0}\right)$	$\dfrac{h_1}{h_0}$	$\phi\left(\dfrac{h_1}{h_0}\right)$	$\phi\left(\dfrac{h_2}{h_0}\right) - \phi\left(\dfrac{h_1}{h_0}\right)$	$f(h_0)$
1.85	2.16	2.31	1.08	-1.45	3.76	6.95
1.9	2.10	2.20	1.05	-1.95	4.15	7.88
1.95	2.05	2.10	1.025	-2.66	4.76	9.29

Plotting the results of Table 2-1 in Fig. 2-11, we obtain the value $h_0 = 1.88$.

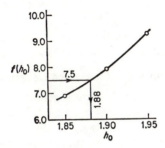

FIG. 2-11

From Eq. (4), Sec. 2-4, we find the quantity of seepage from the canal to the river is

$$Q = 0.002 \times 0.025 \times 188 \times 200,000 \text{ cm}^3/\text{sec} = 0.50 \text{ gal/sec}$$

To find the height of the line of seepage above the impervious boundary it is most convenient to assume values of h ($2.0 < h < 4.0$) and then compute the corresponding distances. Assuming values of $h_2 = 2.5$, 3.0, and 3.5, from Eq. (7), written as

$$L = \frac{1.88}{0.025} \left[\phi\left(\frac{h_2}{1.88}\right) - \phi\left(\frac{2.00}{1.88}\right) \right]$$

we find the respective distances of 144 meters, 210 meters, and 255 meters from the canal as shown in Fig. 2-10.

2-6. Pavlovsky's Solution for $i < 0$

For this case we can have a falling free surface only (Fig. 2-6c). Following the same procedure as in Sec. 2-5, we obtain for the case at hand

$$\frac{iL}{h_0} = \eta_1 - \eta_2 + \ln\frac{\eta_2 + 1}{\eta_1 + 1} \tag{1}$$

Now taking

$$\psi(\eta) = -\eta + \ln(\eta + 1) \tag{2}$$

we see that the form of the solution for this case is

$$\frac{iL}{h_0} = \psi(\eta_2) - \psi(\eta_1) \tag{3}$$

Equation (2) is plotted in Fig. 2-12. The method of problem solving is

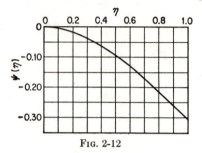

Fig. 2-12

entirely similar to that demonstrated in Example 2-1, except that the free surface lies below the normal depth $(h_0 > h)$.

2-7. Seepage through an Earth Dam on an Impervious Base

In this section we shall consider several solutions for the determination of the discharge and the free surface through homogeneous earth dams on impervious bases. Each of these procedures makes use of Dupuit's assumptions and hence is subject to the limitations outlined in Sec. 2-1. A more rigorous treatment of this problem will be given in a later chapter.

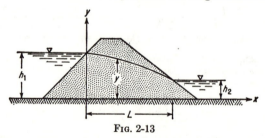

Fig. 2-13

1. Dupuit's Solution. With Dupuit's assumptions, the discharge (per unit width) through any vertical section of the dam in Fig. 2-13 is [Eq. (4), Sec. 2-2]

$$q = -ky\frac{dy}{dx} \tag{1}$$

Integrating and substituting the boundary conditions $x = 0$, $y = h_1$ and $x = L$, $y = h_2$, we obtain Dupuit's formula [Eq. (5), Sec. 2-2]

$$q = \frac{k(h_1^2 - h_2^2)}{2L} \tag{2}$$

Equation (2) specifies a parabolic free surface, commonly referred to as *Dupuit's parabola*. In the derivation above no cognizance has been taken of the entrance or exit conditions (cf. Fig. 1-13) of the line of seepage or of the development of a surface of seepage. Indeed, in the absence of tail water ($h_2 = 0$), the line of seepage is seen to intersect the impervious base. Also, it should be noted that both the discharge quantity and the locus of the free surface are independent of the slopes of the dam.

2. Solution of Schaffernak and Van Iterson. The first approximate method that accounts for the development of the surface of seepage was proposed independently in 1916 by Schaffernak [125] and van Iterson [62].

Considering an earth dam on an impervious base (Fig. 2-14a) with no tail water and applying Eq. (1) to triangle CAB, we obtain for the dis-

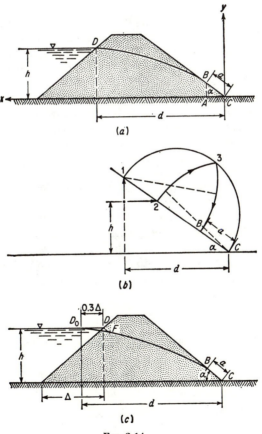

Fig. 2-14

charge per unit width (with x taken as positive to the left).

$$q = ky\frac{dy}{dx} = ka \sin \alpha \tan \alpha \tag{3}$$

where a is the length of the surface of seepage. To determine the value of a, we have from Eq. (3),

$$\int_{a \sin \alpha}^{h} y\, dy = a \sin \alpha \tan \alpha \int_{a \cos \alpha}^{d} dx$$

which, after integration, yields

$$a = \frac{d}{\cos \alpha} - \sqrt{\frac{d^2}{\cos^2 \alpha} - \frac{h^2}{\sin^2 \alpha}} \tag{4}$$

Equation (4) lends itself to a simple graphical construction for the determination of a, as shown in Fig. 2-14b. The known entrance point (d,h) is projected vertically and horizontally to meet the extension of the downstream slope at points 1 and 2. A semicircle with its center on the slope is drawn through points 1 and C, and point 3 is located by striking an arc of radius 2-C with C as a center. Finally, an arc of radius 1-3, with 1 as center, locates point B on the slope and hence the required distance a.

Unlike Dupuit's solution the parabolic free surface for this case is tangent to the downstream slope, as is required (cf. Fig. 1-13b). For the entrance-condition correction at the upstream slope, A. Casagrande [14] recommended that point D_0 (Fig. 2-14c) instead of point D be taken as the starting point of the line of seepage (D_0 is 0.3Δ from point D at the upstream reservoir surface). The actual entrance condition is then obtained by sketching in the arc DF normal to the upstream slope and tangent to the parabolic free surface.

3. L. Casagrande's Solution. Taking exception to Dupuit's second assumption that the hydraulic gradient is equal to the slope dy/dx of the free surface, L. Casagrande [16] analyzed the same problem as Schaffernak and van Iterson with the hydraulic gradient equal to dy/ds, where s is measured along the free surface. Hence, Eq. (1) for Casagrande's method is

$$q = -ky\frac{dy}{ds} \tag{5}$$

Applying Eq. (5) at AB (Fig. 2-15a) we have, for the quantity of seepage,

$$q = ka \sin^2 \alpha \tag{6}$$

Equating the right sides of Eqs. (5) and (6) and setting the limits of integration, we obtain

$$-\int_{h}^{a \sin \alpha} y\, dy = a \sin^2 \alpha \int_{0}^{S-a} ds \tag{7}$$

where S is the length of the line of seepage as shown in Fig. 2-15a.* The solution of Eq. (7) yields

$$a = S - \sqrt{S^2 - \frac{h^2}{\sin^2 \alpha}} \qquad (8)$$

The distance S in Eq. (8) differs but little from the straight line CD_0 of Fig. 2-15a, which may be used as a first approximation. In the event

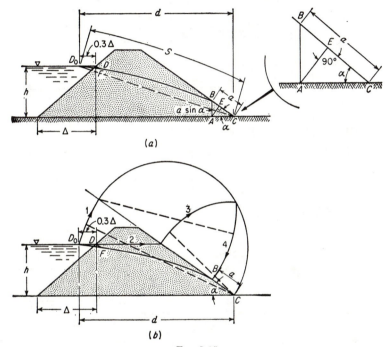

(a)

(b)

Fig. 2-15

that greater precision is required for S, after determining a on the basis of $S = CD_0$, assume as a second approximation $S = a + BD_0$, etc. In other than extraordinary circumstances the first approximation is sufficient; hence we may take

$$S = \sqrt{h^2 + d^2} \qquad (9)$$

and

$$a = \sqrt{d^2 + h^2} - \sqrt{d^2 - h^2 \cot^2 \alpha} \qquad (10)$$

Casagrande recommended a graphical solution of Eq. (10) (Fig. 2-15b) with a construction similar to that given previously in Fig. 2-14b. Once a

* The entrance point is again located at 0.3Δ and a suitable correction is effected by sketching in the curve DF.

has been determined, the quantity of seepage is merely k times BE in Fig. 2-15a.

A solution of Eq. (8) which avoids the approximation of Eq. (9) was obtained by Gilboy [42] in 1933. A modified form of this solution is presented in Fig. 2-16.

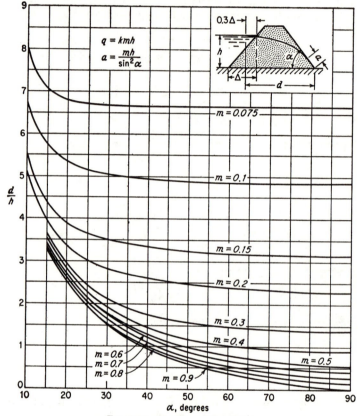

Fig. 2-16. (Based on Gilboy [42].)

Example 2-2. Determine the quantity of seepage for 100 ft of the dam section shown in Fig. 2-17. $k = 0.002$ ft/min.

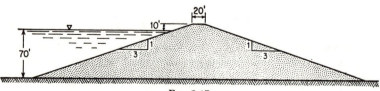

Fig. 2-17

For this case, $\alpha = \tan^{-1} \frac{1}{3} = 18.5°$, $0.3\Delta = 0.3 \times 210 = 63$ ft, and $d = 63 + 3 \times 10 + 20 + 3 \times 80 = 353$. Hence, in Fig. 2-16, with $d/h = {}^{353}\!/_{70} = 5.0$ and $\alpha = 18.5$, we find $m = 0.11$ and the quantity of seepage per 100 ft is

$$Q = 100q = 100kmh = 100 \times 2 \times 10^{-3} \times 0.11 \times 70 = 11.5 \text{ gal/min}$$

The length of the surface of seepage for this case is $a = mh/\sin^2 \alpha = 77$ ft.

4. Pavlovsky's Solution.   Pavlovsky [108] considered the dam to be divided into three zones, as shown in Fig. 2-18.   He assumed the upper

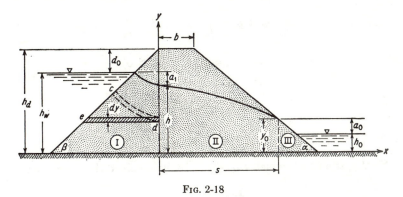

Fig. 2-18

section (I) to be bounded by the upstream slope and the y axis,* the central section (II) by the y axis and a vertical line through the discharge point of the free surface, and the lower section (III) by the latter vertical line and the downstream slope.

The streamlines in zone I are known to be curvilinear (dotted curves cd in Fig. 2-18); however, Pavlovsky assumed that they may be replaced by horizontal streamlines of almost equivalent length ed. Then, assuming purely horizontal flow in zone I, the discharge through an elemental strip is

$$dq = k \frac{a_1}{m(h_d - y)} dy \tag{11}$$

where $m = \cot \beta$ and $a_1/[m(h_d - y)]$ is the hydraulic gradient. Now, setting the limits of integration

$$q = k \frac{a_1}{m} \int_0^h \frac{dy}{h_d - y}$$

he obtained for the reduced quantity of seepage in zone I

$$\frac{q}{k} = \frac{h_w - h}{m} \ln \frac{h_d}{h_d - h} \tag{12}$$

* See Prob. 19.

For zone II, Pavlovsky used Dupuit's formula [Eq. (5), Sec. 2-2],

$$\frac{q}{k} = \frac{h^2 - (a_0 + h_0)^2}{2s} \tag{13}$$

Two conditions must be considered for zone III, depending on whether or not tail water is present. Here again the assumption of horizontal flow is made. If tail water is absent ($h_0 = 0$),

$$\int_0^q dq = k \frac{1}{m_1} \int_0^{a_0} dy \qquad \frac{q}{k} = \frac{a_0}{m_1} \tag{14}$$

where $m_1 = \cot \alpha$. Considering tail water ($h_0 > 0$)

$$\frac{q}{k} = \int_{h_0}^{y_0} \frac{dy}{m_1} + \int_0^{h_0} \frac{y_0 - h_0}{m_1(y_0 - y)} \, dy$$

and

$$\frac{q}{k} = \frac{a_0}{m_1} \left(1 + \ln \frac{a_0 + h_0}{a_0} \right) \tag{15}$$

Finally, from the geometry of Fig. 2-18, he obtained the expression

$$s = b + m_1[h_d - (a_0 + h_0)] \tag{16}$$

The four independent equations [Eqs. (12), (13), (15), and (16)] contain only the four unknowns, h, a_0, s, and q, and hence provide a complete solution.

If tail water can be neglected ($h_0 = 0$), combining Eqs. (13) and (14) and substituting Eq. (16) for s, we obtain

$$a_0 = \frac{b}{m_1} + h_d - \sqrt{\left(\frac{b}{m_1} + h_d\right)^2 - h^2} \tag{17}$$

Likewise, from Eqs. (12) and (14),

$$\frac{a_0 m}{m_1} = (h_w - h) \ln \frac{h_d}{h_d - h} \tag{18}$$

Now, Eqs. (17) and (18) contain only two unknowns (a_0 and h) and hence can be solved without difficulty.

Example 2-3. Determine the quantity of seepage, a_0, and h using Pavlovsky's solution for the dam of Example 2-2.

For this case, $h_d = 80$ ft, $b = 20$ ft, $m = m_1 = 3$, and $h_w = 70$ ft. For assumed values of h in Eqs. (17) and (18), the resulting values for a_0 are given in Table 2-2 and plotted in Fig. 2-19. Thus, $a_0 = 18.3$ and $h = 52.7$ ft, and from Eq. (14)

$$Q = 100q = 100 \times 2 \times 10^{-3} \times \frac{18.3}{3} = 9.1 \text{ gal/min}$$

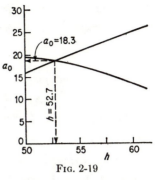

FIG. 2-19

Table 2-2

Equation (17)		Equation (18)	
h	a_0	h	a_0
50	16.0	50	19.6
52.5	18.1	52.5	18.7
55	20.1	55	17.4
60	24.4	60	13.9
65	29.6	65	8.4

The length of the horizontal projection of the surface of seepage for this case is $3a_0 = 54.9$ ft.

2-8. Radial Flow into Completely Penetrating Wells

The quantity of seepage at any distance r from the center of the well in Fig. 2-20a is

$$Q = kiA \tag{1}$$

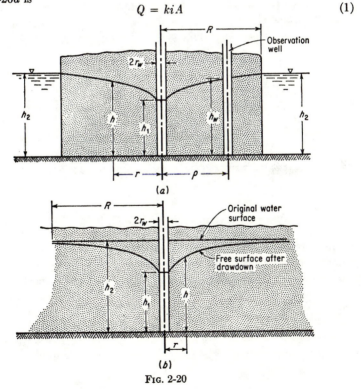

FIG. 2-20

Noting that the area A through which flow occurs is $2\pi rh$ and applying Dupuit's assumptions ($i = dh/dr$), we obtain

$$Q = k \, 2\pi rh \frac{dh}{dr} \tag{2}$$

which, setting the limits of integration, yields

$$\frac{Q}{2\pi k} \int_{r_w}^{R} \frac{dr}{r} = \int_{h_1}^{h_2} h \, dh$$

and

$$Q = \pi k (h_2{}^2 - h_1{}^2) \frac{1}{\ln (R/r_w)} \tag{3}$$

We note at this point that the above derivation applies equally as well to the condition depicted in Fig. 2-20b. Here it is assumed that the surface is tangent to the original ground surface at some radial distance R, called the *radius of influence of the well*. Noting that the quantity of seepage is rather insensitive to the ratio R/r_w (see Prob. 20), Muskat [99] observed that any reasonable assumption for R will provide sufficiently precise estimates for Q.

Equation (3) was first derived by Dupuit in 1863 and, although it neglects completely the development of a surface of seepage at the well (Fig. 2-21), the quantity of flow obtained from this expression has since

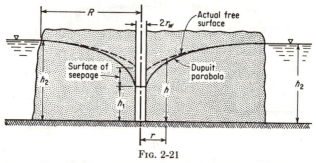

Fig. 2-21

been corroborated by several investigators [5, 99]. However, the location of the free surface predicted by the Dupuit theory in the general vicinity of the well has been less satisfactory. This is due in a large measure to the omission of the surface of seepage (Fig. 2-21). Based upon extensive investigations of this problem with sand models and electrical models, Babbitt and Caldwell [5] concluded that the shape of the free surface closely approximates the Dupuit curve at distances greater than h_2 from the well. Thus, if an observation well is located at a radial distance ρ (Fig. 2-20a), where $\rho > h_2$, the height of the free surface above the impervious boundary h at any $r > h_2$ can be determined from

$$h^2 = \frac{h_2{}^2 - h_1{}^2}{\ln (R/r_w)} \ln \frac{r}{\rho} + h_w{}^2 \tag{4}$$

where h_w is the elevation of the free surface at the observation well (above the impervious boundary). For $r < h_2$, Babbitt and Caldwell recommended that the height h be obtained from the empirical expression (Fig. 2-21)

$$h = h_2 - \frac{C_x}{h_2} \frac{h_2{}^2 - h_1{}^2}{\ln (R/r_w)} \ln \frac{R}{0.1h_2} \qquad (5)$$

where C_x is a correction factor (Fig. 2-22) that is dependent on the ratio

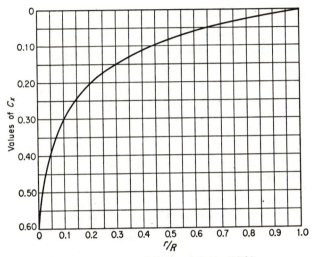

Fig. 2-22. (*After Babbitt and Caldwell* [5].)

of r/R. Recent large-scale tests [47] tend to substantiate this approximation.

Problems of partially penetrating wells, eccentric wells, interference between wells, and the like will be investigated in some detail in a later chapter.

PROBLEMS

1. Obtain the third term of Taylor's series in Eq. (2), Sec. 2-1.

2. Two observation wells are located as shown in Fig. 2-23. The observed elevations of the free surface above the horizontal impervious boundary are $h_1 = 40.3$ ft and $h_2 = 35.5$ ft. The coefficient of permeability is 30 ft/day. Determine (*a*) the elevation of the free surface at section M; (*b*) the quantity of seepage per 10 ft of section; (*c*) the distribution of the horizontal and vertical velocities at section M.

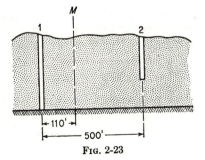

Fig. 2-23

3. Demonstrate that Dupuit's formula can be obtained directly (without integration) from $q = kiA$, where A is the area normal to the direction of flow and i is the hydraulic gradient.

4. Complete the intermediate steps between Eqs. (6) and (7), Sec. 2-2.

5. The section shown in Fig. 2-24 reaches a steady state of flow under a uniform infiltration of e. Find the ratio e/k and construct the free surface for the section if the

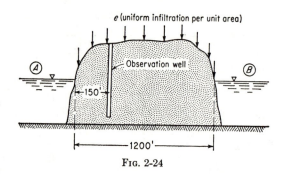

e (uniform infiltration per unit area)

Observation well

Ⓐ

Ⓑ

←150'

←————1200'————→

Fig. 2-24

water level above the impervious base in rivers A and B is 20 ft and the elevation of the water surface in the observation well is 21 ft.

6. In the case of uniform infiltration onto a free surface between two rivers, demonstrate that the water divide is closer to the river with the higher water level.

7. Placing the origin of coordinates at $a = L/2$ as shown in Fig. 2-25 and assuming $h_{max} = b$, demonstrate that the equation for the free surface is the ellipse $h^2/a^2 + x^2/b^2 = 1$.

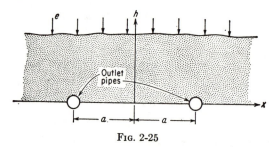

e

h

Outlet pipes

←——a——→←——a——→

x

Fig. 2-25

8. If h_0, h_1, and h are such that the ratio $(h_1 - h_0)/(h - h_0)$ is close to unity, demonstrate that Eq. (2), Sec. 2-5, is approximately

$$h = h_1 - i(x - x_1) + 2h_0 \frac{h_1 - h}{h + h_1 - 2h_0}$$

9. Using the results of Prob. 8, demonstrate that the general Dupuit formula for an inclined impervious boundary ($i = \tan \alpha$) is

$$q = -k \frac{h_1 + h_2}{2} \left(\frac{h_2 - h_1}{x_2 - x_1} + i \right)$$

10. The section shown in Fig. 2-26 consists of a sand with $k = 44.0$ ft/day. The water level at the river is at an elevation of 128.8 ft, 14 ft above an inclined layer of

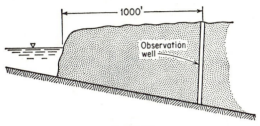

FIG. 2-26

relatively impervious clay. The water level at the observation well is at 140.0 ft, 32 ft above the clay layer. Determine (a) the normal depth and (b) the discharge per 10 ft of section; and (c) construct the free surface between the river and the well.

11. Repeat Prob. 10 with $k = 4.4$ ft/day.

12. Solve Example 2-1 if the depth of water in the canal (h_1) is 4 meters and the depth of water in the river (h_2) is 2 meters.

13. Using the Dupuit analysis for flow through an earth dam and neglecting tail water, determine the velocity and the slope of the free surface at the point of discharge. The impervious boundary is horizontal.

14. Verify the graphical constructions for the determination of the surface of seepage for (a) the Schaffernak and van Iterson solution and (b) L. Casagrande's solution.

15. Demonstrate the equivalence of Eqs. (3) and (14), Sec. 2-7.

16. Solve Example 2-2 by the methods of Dupuit and Schaffernak and van Iterson.

17. Solve Example 2-2 by each of the approximate methods if the downstream slope in Fig. 2-17 is 1 horizontal to 3 vertical.

18. Draw a flow net for the dam in Prob. 17 and evaluate the various approximate solutions.

19. In Fig. 2-17, using Pavlovsky's solution and choosing several positions for the location of the y axis, obtain a plot of the reduced quantity of seepage and a_0 as a function of the position of the y axis.

20. In Fig. 2-20b, $r_w = 0.5$ ft and $R = 100$ ft. Find the allowable variation in R so that the error in the quantity of seepage is within ± 10 per cent.

21. Obtain the free surface for the well in Fig. 2-27 by (a) Dupuit's theory and (b) the Babbitt and Caldwell recommendations.

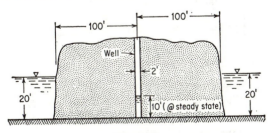

FIG. 2-27

3

Conformal Mapping by Elementary Functions

3-1. Introduction and Geometrical Representation of $w = f(z)$

Much of the analytical method for the solution of two-dimensional groundwater problems is concerned with the determination of a function which will transform a problem from a geometrical domain within which a solution is sought into one within which the solution is known. In this chapter we shall study various elementary functions and the manner in which these functions transform geometric figures from one complex plane to another. In Chap. 4 we shall consider the more general problem of finding a functional relationship that will provide a specific transformation. It will be assumed that the reader has attained a working knowledge of the elements of complex variable theory (see Appendix A).

Let $w = \phi + i\psi$ be an analytic function of $z = x + iy$, and suppose that a complex number $x + iy$ is located at point P_1 in the z plane (Fig. 3-1a). As w is a function of z there must be some point Q_1 in the w plane

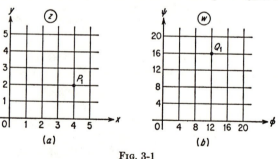

Fig. 3-1

(Fig. 3-1b) corresponding to the point P_1 in the z plane. It is desirable at this point to illustrate this relationship by an example.

Example 3-1. If $w = z^2$, determine the point in the w plane corresponding to the point $z = 4 + 2i$ in the z plane (P_1 of Fig. 3-1a).

Expanding $w = z^2$,

$$w = \phi + i\psi = (x + iy)^2 = x^2 + 2ixy - y^2$$

and equating real and imaginary parts, we get

$$\phi = x^2 - y^2 \qquad \psi = 2xy$$

By substitution of $x = 4$, $y = 2$, the corresponding point in the w plane is located at $w = 12 + 16i$, point Q_1 of Fig. 3-1b.

Similarly, by the correspondence of a sequence of points, for any curve in the z plane there will be a corresponding curve in the w plane (subject to certain limitations which will be developed in Sec. 3-3). This procedure is called *mapping*.

Example 3-2. Discuss the mapping of the w plane into the z plane by the function $z = w^2$ and locate in the z plane the area bounded by the lines $\psi = \frac{1}{2}$, $\psi = 1$, $\phi = \frac{1}{2}$, and $\phi = 1$ in the w plane (Fig. 3-2a).

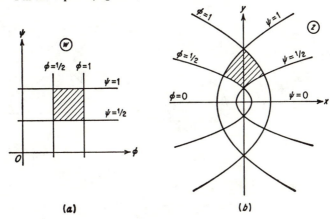

(a)   **(b)**

Fig. 3-2. $z = w^2$.

In this case $x + iy = (\phi + i\psi)^2$, and the parametric equations $x = \phi^2 - \psi^2$ and $y = 2\phi\psi$ provide the transformation between points in the two planes. Now, considering the mapping into the z plane of lines parallel to the ψ axis in the w plane ($\phi = c_1$), we have the parametric equations

$$y = 2c_1\psi \qquad x = c_1^2 - \psi^2$$

Eliminating the parameter ψ, we obtain

$$x = c_1^2 - \frac{y^2}{4c_1^2}$$

which defines a family of parabolas symmetrical about the x axis with focus at the origin of the z plane and opening to the left. Similarly, for lines parallel to the ϕ axis ($\psi = c_2$),

$$x = \frac{y^2}{4c_2^2} - c_2^2$$

which is the equation of a family of parabolas, confocal to those for $\phi = c_1$ but opening to the right.

The confocal parabolas for $\phi = \frac{1}{2}$, $\phi = 1$, $\psi = \frac{1}{2}$, and $\psi = 1$ are presented in Fig. 3-2b. The required area is shown crosshatched.

If ψ in Fig. 3-2a is taken as a stream function and ϕ as the corresponding velocity potential, the function $z = w^2$ transforms a pattern of uniform flow in the w plane into a flow pattern of confocal parabolas in the z plane. This is the starting point of Kozeny's solution for the seepage through an earth dam on an impervious base.

3-2. Application of the Mapping Function $z = w^2$

Kozeny [76] studied the problem of seepage through an earth dam with a parabolic upstream face resting on an impervious base (Fig. 3-3a).

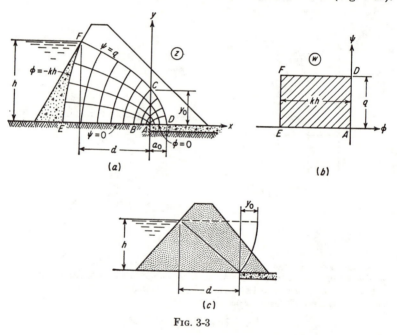

(a)

(b)

(c)

Fig. 3-3

In addition, an horizontal toe drain (underfilter) was located at the downstream portion of the dam. The purpose of the drain is to control seepage through the dam. The discharge from the drain is generally collected by pipes (Fig. 3-4a) and led into the spillway stilling basin or into the river channel below the dam. Soils used between the dam proper and the pipe (or pipes) are designed as graded filters (see Art. 11, Ref. 145 and Sec. 126, Ref. 148). Two other types of toe drains in

common usage are the trapezoidal
toe drain (Fig. 3-4b) and the slope
drain (Fig. 3-4c).

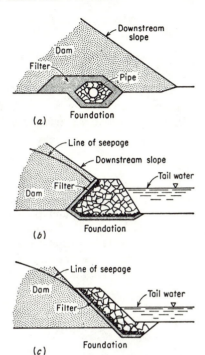

Whereas the coefficient of permeability of the underdrain in Fig.
3-3a is very much greater than that
of the dam proper, from the standpoint of analysis the boundary between the dam and the drain may
be taken as an equipotential line.
Thus in Fig. 3-3a the region of flow
will be bounded by equipotential
surfaces at the parabolic upstream
face of the dam and along the
boundary of the horizontal drain
AD, and by the streamlines at the
free surface FCD and the impervious
boundary AE. With $\phi = 0$ along
AD, $\phi = -kh$ along EF (where h is
the total head loss through the
dam), $\psi = 0$ along AE, and $\psi = q$
[Eq. (3), Sec. 1-10] at the free surface, the region of flow in the w
plane is as shown in Fig. 3-3b.
Kozeny recognized that the required correspondence between the z plane (Fig. 3-3a) and the w plane
will be provided by the function $z = Cw^2$, where C is constant. Consequently, along the free surface where $\phi = -ky$ and $\psi = q$,

Fig. 3-4

$$x = C(k^2y^2 - q^2) \qquad y = -2Ckyq \tag{1}$$

From the second of Eqs. (1),

$$C = -\frac{1}{2kq}$$

and the equation of the free surface, called *Kozeny's basic parabola*, is

$$x = -\frac{ky^2}{2q} + \frac{q}{2k} \tag{2}$$

Setting $x = 0$ in Eq. (2) and calling the y intercept of the free surface
y_0, the quantity of seepage per unit length of dam is seen to be

$$q = ky_0 \tag{3}$$

Combining Eqs. (2) and (3),

$$y^2 - y_0^2 + 2y_0x = 0 \tag{4}$$

and solving for y_0, we find

$$y_0 = x \pm \sqrt{x^2 + y^2} \qquad (5)$$

This, after the substitution of the entrance coordinates $x = -d$, $y = h$ (Fig. 3-3a), yields

$$y_0 = \sqrt{d^2 + h^2} - d \qquad (6)$$

A graphical solution of Eq. (6) is shown in Fig. 3-3c. If $y = 0$ in Eq. (2), the focal distance (also called the *minimum length of underdrain*) equals

$$a_0 = \frac{y_0}{2} \qquad (7)$$

Differentiation of Eq. (4) with respect to y shows the slope at any point of the free surface to be $-y_0/y$. Hence, the free surface enters the drain vertically ($y = 0$) and its slope at $y = y_0$ is $-45°$. Thus all required information is available for the complete solution of the problem.

A. Casagrande [14] extended Kozeny's solution to include dams with trapezoidal toe drains and slope drains. He begins by drawing Kozeny's basic parabola with point A (Fig. 3-5a) as the focus and passing through

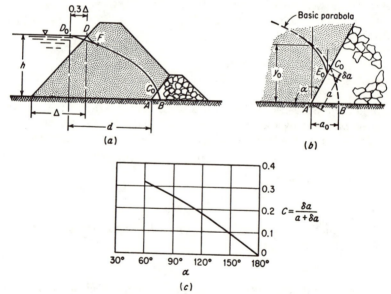

Fig. 3-5. (*After Casagrande [14].*)

the points D_0, C_0, and B. The location of the entrance point D_0 at 0.3Δ from point D is suggested as an average value to allow for the deviation of the upstream slope from the parabolic face assumed in Kozeny's solution. The entrance condition is adjusted by sketching in the arc DF normal to the upstream slope and tangent to the parabolic free surface (see Sec. 2-7). By constructing flow nets in the vicinity of inclined toe drains (with varying slopes α), Casagrande found the distance δa between the point C_0 on the basic parabola and the point E_0 on the line of seepage (Fig. 3-5b).   The ratio

$$C = \frac{\delta a}{a + \delta a} \tag{8}$$

is shown plotted as a function of α in Fig. 3-5c.   If it is recognized that the line of seepage enters the sloping drain vertically (Fig. 1-13b), its exit portion can be sketched in with a fair degree of approximation.   The discharge per unit length of dam may be determined with sufficient accuracy [Eq. (3)] from

$$q = k(\sqrt{d^2 + h^2} - d) \tag{9}$$

Figure 3-6a represents the downstream portion of an earth dam with tail water. AB is a surface of seepage and BE is an equipotential line.

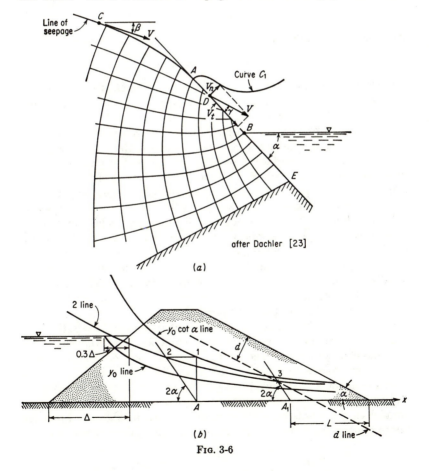

after Dachler [23]

(a)

(b)

FIG. 3-6

Thus, at point A the line of seepage must enter tangent to the discharge slope, whereas streamlines entering below point B must intersect the slope at right angles. Calling γ the angle at which a streamline intersects the surface of seepage, we see that γ varies from zero at point A to 90° at point B. At point C on the line of seepage the resultant velocity V forms an angle β with the horizontal. Denoting the direction of the free surface as s,

$$V = -k\frac{dh}{ds} = -k \sin \beta \tag{10}$$

At point A, the discharge point, the line of seepage enters tangent to the discharge slope and hence the normal component of the velocity V_n (normal to the discharge slope) must be zero and the tangential component of the velocity is

$$V_t = V = -k \sin \alpha \tag{11}$$

Along the discharge slope (cf. point D) the normal component is

$$V_n = -k \sin \alpha \tan \gamma \tag{12}$$

Thus, with increasing values of γ, the normal component of the velocity increases with $\tan \gamma$, and at point B where $\gamma = 90°$ theoretically, the normal component of the velocity (and hence the exit gradient) is unbounded (curve C_1). In practice, of course, infinite velocities would not develop (Darcy's law would no longer be valid); however, the high velocities in the region where the surface of seepage approaches tail water can result in severe downstream erosion. An important function of the toe drains (Fig. 3-4) is to avoid the formation of a surface of seepage along discharge slopes, thereby eliminating the possibility of a serious erosion problem.

Inasmuch as toe drainage tends to pull the line of seepage in from the downstream slope, a horizontal toe filter can be located so as to prevent the line of seepage from coming within a specified distance from the downstream slope. Such a method will be developed graphically, making use of Kozeny's basic parabola with the 0.3Δ entrance modification. The proof of this method can be obtained by comparing the specified constructions with those of the basic parabola.

In Fig. 3-6b, with various positions of point A along the x axis, the locus of points y_0, the y intercept of the basic parabola, is determined. Next the locus of $y_0 \cot \alpha$ is obtained. From points A (only one construction is shown) are constructed lines at an angle 2α. Then horizontal lines from the $y_0 \cot \alpha$ line above points A (point 1) are intersected with the 2α line (at point 2), yielding the locus of points 2. A line is then drawn parallel to the downstream slope at the desired distance d. From the intersection of the d line with the 2 line (point 3), a line is drawn at 2α to intersect the x axis at A_1. The distance L is the minimum length of underdrain that would prevent any part of the line of seepage from coming closer than a distance d to the downstream slope. Graphical solutions indicate that the procedure will also yield valid results for trapezoidal drains with L taken as the base of the trapezoidal section.

3-3. Conformal Mapping

In the preceding sections it was noted that the transformation $z = w^2$ provided a certain geometrical similarity between the z plane and the w plane; that is, the mapping function preserved the angles of intersection and the approximate geometric shapes between planes except at the origin. A transformation that possesses the property of preserving angles of intersection and the approximate image of small shapes is said to be *conformal*. In the present section we shall explain the behavior of conformal transformations and the conditions under which they fail.

In Sec. A-3 it is shown that if $w = \phi + i\psi = f(z)$ is analytic within a region R its derivative $f'(z)$ is single-valued; that is, $f'(z)$ has only one value at any point z in R. However, as z varies from point to point, $f'(z)$ will in general be a function of z. In Fig. 3-7, let C be a smooth curve

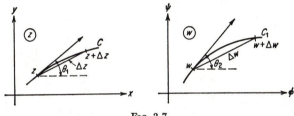

FIG. 3-7

through a point z, and let C_1 be its image through point w under the transformation $w = f(z)$ when $f(z)$ is analytic at z and $f'(z) \neq 0$. As $f'(z)$ must be a complex number, say $f'(z) = A \exp i\alpha$, then from the definition of a derivative

$$f'(z) = \frac{dw}{dz} = \lim_{\Delta z \to 0} \frac{\Delta w}{\Delta z}$$

we obtain the two equations

$$\alpha = \arg f'(z) = \lim_{\Delta z \to 0} \left(\arg \frac{\Delta w}{\Delta z} \right) \tag{1}$$

$$A = |f'(z)| = \lim_{\Delta z \to 0} \left| \frac{\Delta w}{\Delta z} \right| \tag{2}$$

Now, as $\Delta z \to 0$, the limit of the argument of Δz approaches the angle θ_1. In a similar manner, as Δw is the image of Δz, the argument of Δw

approaches the angle θ_2 as $\Delta z \to 0$.   Hence from Eq. (1)

$$\alpha = \arg f'(z) = \theta_2 - \theta_1$$

or

$$\theta_2 = \alpha + \theta_1 \tag{3}$$

Thus in the transformation from the z plane to the w plane the directed tangent to a curve at point z is rotated through the angle $\alpha = \arg f'(z)$. Now, as $f'(z)$ has only one value at any point z, any two curves intersecting at a particular angle at point z will, even after transformation, intersect at the same angle at w (the image of z); that is, the sides of the angle at w are rotated in the same direction by the same amount.

Similarly, as $\Delta z \to 0$, we conclude from Eq. (2) that in the transformation from the z plane to the w plane, infinitesimal lengths in z are magnified at w by the factor $A = \mod f'(z)$.   Now as $\Delta z \to \varepsilon > 0$, Eq. (2) becomes only approximate and hence there is some distortion in the length Δw, with the degree of distortion depending on the magnitude of ε.   Thus large figures in the z plane may transform into shapes bearing little resemblance to the original but, it should be emphasized, the angles formed by corresponding intersecting curves in these planes are preserved exactly even for large figures (except where $f'(z) = 0$).

Points at which $f'(z) = 0$ are said to be *critical points* of the transformation; that is, they represent points where angles are not preserved conformally.   For example, the transformation $w = z^2$ (the nature of this mapping can be inferred by interchanging the z and w planes in Fig. 3-2) demonstrates that angles at the origin where $f'(z) = 0$ are doubled. For the function $w = z^3$, angles at the origin are tripled.   Indeed, it can be shown that for $w = z^n$, angles at $z = 0$ are multiplied $n$fold.   A function $f'(z)$ is said to have $n$fold zeros when n is the number of derivatives which are zero for a particular z.   In other words, for $w = z^2$ the derivative $f'(z)$ has one zero for $z = 0$; $w = z^3$ has a double zero at the origin.   Thus we can generalize that if $f'(z)$ has $n$fold zeros, angles are not preserved at the critical points but are multiplied $(n + 1)$ times (one greater than the number of zero derivatives).

3-4. Fundamentals of Solution of Two-dimensional Flow Problems by Conformal Mapping

The usefulness of conformal mapping in two-dimensional flow problems stems from the fact that solutions of Laplace's equation remain solutions when subjected to conformal transformations.

Let $w = \phi + i\psi = f(z)$ be the complex potential and let its real and imaginary parts satisfy Laplace's equation in the region R of the z plane (Fig. 3-8a), so that

$$\frac{\partial^2 \phi}{\partial x^2} + \frac{\partial^2 \phi}{\partial y^2} = 0 \qquad \frac{\partial^2 \psi}{\partial x^2} + \frac{\partial^2 \psi}{\partial y^2} = 0$$

Now suppose that there is a second analytic function $z = F(t)$, with $t = r + is$, which maps the interior of the curve C into the interior of the

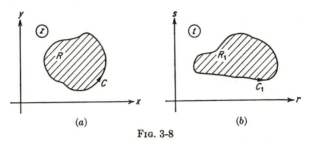

Fig. 3-8

curve C_1 (Fig. 3-8b). The function $w = f[F(t)]$ is an analytic function of an analytic function, which in turn is also analytic, and hence

$$\frac{\partial^2 \phi}{\partial r^2} + \frac{\partial^2 \phi}{\partial s^2} = 0 \qquad \frac{\partial^2 \psi}{\partial r^2} + \frac{\partial^2 \psi}{\partial s^2} = 0$$

In Chap. 1 it was shown that the solution of a two-dimensional ground-water problem could be reduced to one of seeking the solution of Laplace's equation subject to certain boundary conditions within a region R in the z plane. A more or less direct attack was provided by the graphical construction of flow nets (Sec. 1-12). From the standpoint of an analytical solution to Laplace's equation, unless the region R is of a very simple shape a direct approach to the problem is generally very difficult. However, by means of conformal mapping, it is often possible to transform the region R into a simpler region R_1 wherein Laplace's equation can be solved subject to the transformed boundary conditions. (For example, ϕ and ψ are solutions of Laplace's equation and are known at all points of the w plane in Fig. 3-2.) Once the solution has been obtained in region R_1, it can be carried back by the inverse transformation to the region R, the original problem. Hence the crux of the problem is finding a transformation (or series of transformations) that will map a region R conformally into a region R_1 so that R_1 will be of a simple shape, such as a rectangle (whose sides may even extend to infinity) or a circle.

To provide a catalogue of such functions we shall now consider some simple yet powerful classes of conformal transformations.

3-5. Linear Mapping Function

The general form of the linear mapping function is

$$w = az + b \qquad a \neq 0 \tag{1}$$

where a and b are any complex constants. This mapping is everywhere conformal, since $f'(z) = a \neq 0$.

If $a = 1$, $w = z + b$, and each point in w is found by translating each point z through the vector b (Fig. 3-9).

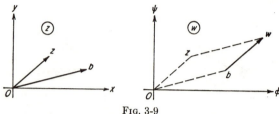

Fig. 3-9

If $|a| = 1$, say $a = \exp i\alpha$, and $b = 0$ then in Eq. (1) $|w| = |z|$ and $\arg w = \arg z + \alpha$. Thus the transformation represents a rotation of each z through the angle α (Fig. 3-10).

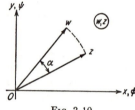

Fig. 3-10

If $b = 0$ and $a = r$, where r is a real number, then $\arg w = \arg z$, $|w| = r|z|$, and the mapping consists of a multiplication of a radius vector in z by the factor r (Fig. 3-11).

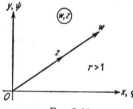

Fig. 3-11

Combining the three cases just described, we can conclude that the general linear transformation $w = az + b$ represents a rotation through the angle $\arg a$, a multiplication by the factor $|a|$, and then a translation through the vector b. In the transformation a region in the z plane is transformed into a geometrically similar region in the w plane.

3-6. Reciprocal Function, $w = 1/z$

The transformation $w = 1/z$ provides a single-valued correspondence between all points in the z and w planes. In particular, points at the origin $z = 0$ (or $w = 0$) are mapped into $w = \infty$ (or $z = \infty$). Taking $w = \rho \exp i\alpha$ and $z = r \exp i\theta$, the transformation yields

$$\alpha = -\theta \qquad \rho = \frac{1}{r}$$

The first equality implies a simple reflection about the real axis; that is, $\arg w = - \arg z$ (Fig. 3-12). This is followed by an *inversion*

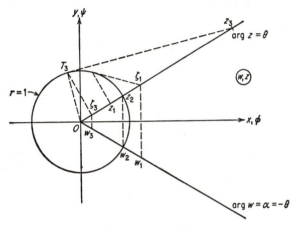

Fig. 3-12. $w = 1/z$.

with respect to the unit circle $r = 1$; $|w||z| = 1$. In the inversion process the points $r = 1$ play a special role in that they define a circle of unit radius in the z plane that maps into itself in the w plane. The points $r < 1$ in the z plane (points interior to the unit circle) correspond to the points $\rho > 1$ (points exterior to the unit circle) in the w plane, or, in general, points exterior to the unit circle are mapped into its interior. In other words, the inversion process can be thought of as a reflection about the unit circle. By a familiar theorem from geometry, a point and its inverse are related as follows: assume that the circle about point O (Fig. 3-12) represents the unit circle ($r = 1$) and that the point z_3 is an exterior point in the z plane ($r > 1$). From z_3 construct the tangent to T_3 and drop a perpendicular to ζ_3 on the line Oz_3. ζ_3 is said to be the *inverse* of the point z_3, and conversely. The validity of the construction is evident from

$$\frac{O\zeta_3}{OT_3} = \frac{OT_3}{Oz_3}$$

hence $O\zeta_3 \cdot Oz_3 = (OT_3)^2 = 1$. Now as $Oz_3 = |z|$ and $O\zeta_3 = |w|$ the corresponding point in the w plane (w_3) is obtained by reflecting ζ_3 about the real axis. The mappings of some points are shown in the figure.

The usefulness of the reciprocal function $w = 1/z$ stems from its property of transforming circles into circles, considering straight lines as degenerate cases of circles. To illustrate, we begin with the equation of a circle in the z plane

$$a(x^2 + y^2) + bx + cy + d = 0 \qquad (1)$$

where a, b, c, and d are real numbers. Noting that the transformation $w = 1/z$ specifies the relations

$$x = \frac{\phi}{\phi^2 + \psi^2} \qquad y = -\frac{\psi}{\phi^2 + \psi^2}$$

$$\phi = \frac{x}{x^2 + y^2} \qquad \psi = -\frac{y}{x^2 + y^2}$$

in the w plane, Eq. (1) transforms into the circle

$$d(\phi^2 + \psi^2) + b\phi - c\psi + a = 0 \qquad (2)$$

If $a = 0$ in Eq. (1), the expression represents a straight line which maps into a circle through the origin [Eq. (2)] under the transformation. Also, any circle passing through the origin [$d = 0$ in Eq. (1)] will be transformed into a straight line [$d = 0$ in Eq. (2)]. Hence, in summary, the transformation $w = 1/z$ takes: (1) circles not passing through the origin into circles not through the origin, (2) circles through the origin into straight lines not through the origin, and (3) straight lines through the origin into straight lines through the origin. These properties of the reciprocal transformation $w = 1/z$ render it particularly valuable when dealing with hodographs (Chap. 4).

Example 3-3. Find the mapping of the lines $x = C_1$ and $y = C_2$ (Fig. 3-13a) under the transformation $w = 1/z$.

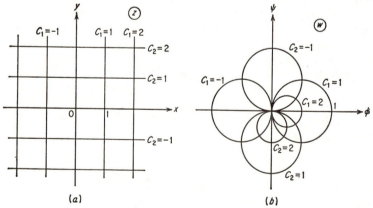

Fig. 3-13. $w = 1/z$.

For the lines $x = C_1$, from Eq. (1), $a = 0$, $c = 0$, and $C_1 = -d/b$. Hence, under the transformation, the lines map into the circles

$$\phi^2 + \psi^2 - \frac{\phi}{C_1} = 0 \tag{3}$$

By the same procedure, the lines $y = C_2$ map into the circles

$$\phi^2 + \psi^2 + \frac{\psi}{C_2} = 0 \tag{4}$$

The required transformation is shown in Fig. 3-13b.

3-7. Bilinear Transformation

The general form of the bilinear transformation is

$$w = \frac{az + b}{cz + d} \qquad ad - bc \neq 0 \tag{1}$$

where a, b, c, and d are complex constants. The requirement $ad - bc \neq 0$ is necessary to ensure the conformal nature of the transformation; that is, if $ad - bc = 0$ every point in the z plane is a critical point. The proof is left as an exercise. In addition, if $a/c = b/d$, w will be a constant irrespective of z and hence the transformation will map the entire z plane into a point in the w plane.

A more convenient form for demonstrating the nature of the transformation is

$$w = \frac{a}{c} + \frac{bc - ad}{(cz + d)c} \tag{2}$$

which, in turn, can be represented by the three successive transformations,

$$\begin{aligned} Z &= cz + d \\ W &= \frac{1}{Z} \\ w &= \frac{a}{c} + \frac{bc - ad}{c} W \end{aligned} \tag{3}$$

The first and third [and $c = 0$ in Eq. (1)] are simple linear transformations of the type $w = Az + B$ considered in Sec. 3-5. The second is the reciprocal function of type $w = 1/z$ investigated in Sec. 3-6. Hence, as each of these transformations maps circles and lines into circles and lines, the bilinear transformation has the same property.

The bilinear transformation has been shown to map circles from one plane into another with straight lines as limiting cases. Thus, as three points in a given plane define a circle, the bilinear transformation should provide a unique mapping of three points z_1, z_2, and z_3 into three points w_1, w_2, and w_3. This can be done by imposing the corresponding conditions directly into Eq. (1) and solving for the constants. However, it will generally be simpler to use the *cross-ratio formula*

$$\frac{(w - w_1)(w_3 - w_2)}{(w - w_2)(w_3 - w_1)} = \frac{(z - z_1)(z_3 - z_2)}{(z - z_2)(z_3 - z_1)} \tag{4}$$

which can be verified without difficulty by noting that

$$w_m - w_n = \frac{(ad - bc)(z_m - z_n)}{(cz_m + d)(cz_n + d)} \tag{5}$$

Example 3-4. What is the bilinear transformation taking the triples $z = -1$, i, 0 into the corresponding points $w = 0$, i, 1?

Setting up the cross ratios, we have

$$\frac{(w - 0)(1 - i)}{(w - i)(1 - 0)} = \frac{(z + 1)(0 - i)}{(z - i)(0 + 1)}$$

which reduces to the desired transformation

$$w = -\frac{z + 1}{z - 1}$$

3-8. Upper Half of z Plane into Unit Circle in w Plane

Any transformation of the type

$$w = e^{i\theta} \frac{z - z_0}{z - \bar{z}_0} \tag{1}$$

where $0 \leqq \theta \leqq 2\pi$ and Im $(z_0) > 0$ will map the upper half of the z plane [Im $(z) \geqq 0$] into the unit circle $|w| \leqq 1$. To verify this transformation, we take the absolute value of both sides of Eq. (1),

$$|w| = |e^{i\theta}| \frac{|z - z_0|}{|z - \bar{z}_0|} \tag{2}$$

which, since $|e^{i\theta}| = 1$ and $|w| \leqq 1$, reduces to the inequality $|z - z_0| \leqq |z - \bar{z}_0|$. Since the points z and z_0 are both in the upper half plane the distance $|z - z_0|$ cannot exceed the distance $|z - \bar{z}_0|$ (Fig. 3-14). As

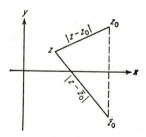

Fig. 3-14

$w = 0$ when $z = z_0$, the center of the unit circle is the image of the point $z = z_0$ (Fig. 3-15).

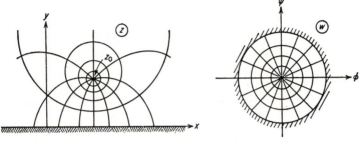

FIG. 3-15

Rather than deal with a point and its conjugate in Eq. (1), let z_0 be a pure imaginary number, say ik; then

$$w = e^{i\theta} \frac{z - ik}{z + ik} \tag{3}$$

or

$$z = ik \frac{e^{i\theta} + w}{e^{i\theta} - w} \tag{4}$$

Now, assuming that on the boundary of the unit circle $w = \exp i\alpha$, from Eq. (4) the correspondence between any point on the unit circle and the real axis of the z plane ($z = x$) is given by

$$x = k \cot \frac{\theta - \alpha}{2} \tag{5}$$

Example 3-5. Demonstrate that the transformation $w = -(z - ik)/(z + ik)$ carries the upper half of the z plane into the unit circle $|w| \leq 1$ in such a way that the first quadrant of the z plane maps into the upper half of the unit circle and the second quadrant of the z plane maps into the lower half of the circle.

Comparing this function with Eq. (3), we see that $\exp i\theta = -1$ and $\theta = \pi$; hence Eq. (5) becomes $x = k \tan (\alpha/2)$. Now, as α varies from 0 to π in the upper half of the unit circle, x is seen to go from 0 to ∞ or along the boundary of the first quadrant of the z plane. Similarly, for $\pi \leq \alpha \leq 2\pi$, x moves along the boundary of the second quadrant. Since the boundaries of the two regions correspond and Eq. (3) takes points in the upper half of the z plane into $|w| \leq 1$, the mapping is verified.

3-9. The Transformation $w = z^n$

If $w = \rho \exp i\alpha$ and $z = r \exp i\theta$, the transformation $w = z^n$ requires that

$$\rho = r^n \qquad \alpha = n\theta \tag{1}$$

Hence in the transformation a sector in the z plane with vertex at the origin and an angle θ will be mapped into a sector in the w plane with vertex at $w = 0$ and a central angle of $\alpha = n\theta$ (Fig. 3-16). The resulting

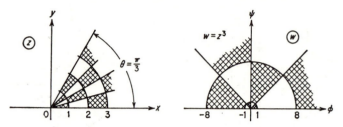

Fig. 3-16

figures will be conformal everywhere except at the origin, where $f'(z) = 0$ (Sec. 3-3).

Example 3-6. Map the upper half of a unit circle in the z plane into the upper half of the w plane (Fig. 3-17).

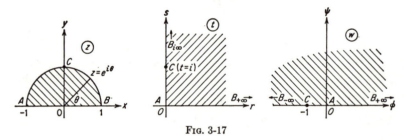

Fig. 3-17

The solution will be obtained in two steps. In the first step the upper half of the unit circle will be mapped into the first quadrant of an auxiliary t plane ($t = r + is$). This will be followed by a mapping of the region in the t plane into the upper half of the w plane.

Let the point $z = -1$ have as its image the point $t = 0$. Using the results of Example 3-5 as

$$t = -\frac{z - ik}{z + ik}$$

would require that $-1 - ik = 0$; hence $k = i$, and the required transformation for the first step is

$$t = -\frac{z + 1}{z - 1} \qquad (2)$$

The second step is performed by opening the sector $\pi/2$ in the t plane into the section π in the w plane. Hence $w = t^2$ and the required solution is

$$w = \left(\frac{z + 1}{z - 1}\right)^2$$

Example 3-7. Find a transformation that will map a sector of angle π/m of a unit circle in the z plane into the upper half of the w plane (Fig. 3-18).

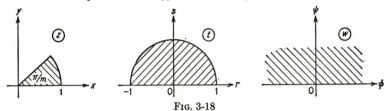

Fig. 3-18

The transformation $t = z^m$ will map the sector into the upper half of a unit circle in the t plane. Therefore, using the results of Example 3-6, the required transformation is

$$w = \left(\frac{z^m + 1}{z^m - 1}\right)^2$$

PROBLEMS

1. Discuss the region into which each of the following regions is mapped under the given transformation. Show the regions graphically: (a) the half plane Im $(z) > 0$ under the transformation $w = z(1 + i)$; (b) the half plane Re $(z) > 0$ under the transformation $w = i(z + 1)$; (c) the area bounded by $\phi = 1$, $\phi = 2$, $\psi = 1$, and $\psi = 2$ under the transformation $w = \ln z$.

2. Discuss the mapping of the lines $\phi = c_1$ and $\psi = c_2$ of the w plane into the z plane under the transformation $w = z^2$. Locate the areas bounded by the lines $\phi = \frac{1}{2}$, $\phi = 1$, $\psi = \frac{1}{2}$, and $\psi = 1$.

3. Solve Example 2-2 with the downstream slope in Fig. 2-17 as 1 horizontal to 3 vertical by A. Casagrande's procedure, and obtain the locus of the free surface.

4. For the dam shown in Fig. 3-19 determine the reduced quantity of seepage and the locus of the free surface.

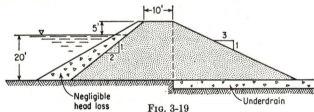

Fig. 3-19

5. For the dam shown in Fig. 3-19 determine the minimum length of the underdrain to prevent any part of the free surface from coming closer than 10 ft to the downstream slope.

6. Repeat Probs. 4 and 5 for the dam section of Fig. 3-20.

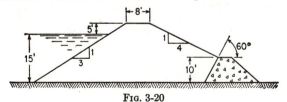

Fig. 3-20

7. Obtain the image of Fig. 3-21 under the transformation $w = (4 + 3i)z + 2$.

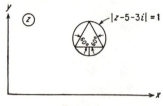

Fig. 3-21

8. Obtain the image in the w plane of the following under the transformation $w = 1/z$:

(a) The region bounded by $x = 5y$, $x < 1$, and $y > 0$
(b) $|z - 3| = 3$
(c) $y = 1$

9. Discuss the transformation $w = z + 1/z$.

10. Find the transformation that will map each of the following triplets in the z plane onto the corresponding points in the w plane:

(a) $z = 0, \infty, 1; \quad w = \infty, 0, i$
(b) $z = 1, i, -1; \quad w = i, -1, -i$
(c) $z = -1, 0, 1; \quad w = -i, 1, i$

11. Obtain the image of Fig. 3-22 under the transformation $w = 1/z$.

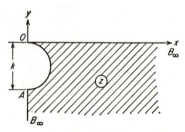

Fig. 3-22

12. Demonstrate that the transformation $w = z/(1 - z)$ maps the interior of the unit circle $|z| < 1$ onto the half plane $\text{Re}(w) > -\frac{1}{2}$.

13. Demonstrate that the transformation $w = \exp z$ takes the infinite strip $-\pi/2 < \text{Im}(z) < \pi/2$ into the right half plane $\text{Re}(w) > 0$.

14. Using the results of Prob. 13, find the transformation that will map the infinite strip $-\pi/4 < \text{Re}(z) < \pi/4$ into the unit circle $|w| < 1$. Hint: First transform the z plane onto the auxiliary t plane by the transformation $t = 2iz$. Why?

4

Special Mapping Techniques

4-1. Velocity Hodograph

Let the complex potential $w = \phi + i\psi$ be an analytic function of the complex variable z, as $w = f(z)$. Differentiating w with respect to z [Eqs. (15) of Appendix A], we find

$$\frac{dw}{dz} = \frac{\partial \phi}{\partial x} + i \frac{\partial \psi}{\partial x}$$

which, substituting the velocity components, yields the *complex velocity*

$$W = \frac{dw}{dz} = u - iv \tag{1}$$

The transformation of the region of flow from the z plane into the W plane is called the *velocity hodograph*. The utility of the hodograph stems from the fact that, although the shape of the free surface and the limit of the surface of seepage are not known initially in the z plane, in the W plane their hodographs are completely defined.

Let us consider the transformation of the various boundaries of a flow region into the W plane.* For simplicity, we shall use the $u + iv$ plane ($\overline{W}$) rather that the $u - iv$ plane (these planes, of course, are merely reflections about the u axis).

1. *Impervious boundary.* At an impervious boundary the velocity vector is in the direction of the boundary. Therefore, designating the boundary at an angle α with the x axis, in the hodograph plane we have

$$\frac{v}{u} = \tan \alpha \tag{2}$$

which represents a straight line in the uv plane, passing through the origin in a direction parallel to the impervious boundary.

2. *Boundary of a reservoir.* The boundary of a reservoir is an equipotential line, $\phi = $ constant; consequently, the velocity vector is per-

---

* We shall not consider singular points here.

pendicular to the boundary. If the equation of the boundary is the straight line $y = x \tan \alpha + b$, then in the hodograph plane the image of the boundary will be the straight line

$$\frac{v}{u} = -\cot \alpha \tag{3}$$

which passes through the origin of the uv plane and is normal to the reservoir boundary.

3. *Free surface.* Along the free surface, $\phi + ky = $ constant. Differentiating this expression with respect to s, where s is along the free surface, and multiplying through by $\partial\phi/\partial s$, we get

$$\left(\frac{\partial\phi}{\partial s}\right)^2 + k\frac{\partial\phi}{\partial s}\frac{dy}{ds} = 0$$

Now, recognizing that $\partial\phi/\partial s$ is the velocity vector, and that

$$\left(\frac{\partial\phi}{\partial s}\right)^2 = u^2 + v^2 \qquad \frac{\partial\phi}{\partial s}\frac{dy}{ds} = v$$

it follows that along the free surface

$$u^2 + v^2 + kv = 0 \tag{4}$$

Equation (4) represents a circle, passing through the origin, with radius $k/2$ and center at $(0, -k/2)$. It follows at once from Eq. (4) that the velocity vector at any point along the free surface can be obtained from the simple graphical construction shown in Fig. 4-1. The angle β in this

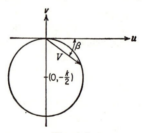

Fig. 4-1

figure represents the angle between the tangent to the free surface at the point in question and the horizontal (see also Fig. 3-6a).

4. *Surface of seepage.* Along the surface of seepage, as in the case of the free surface, $\phi + ky = $ constant. Differentiating this expression with respect to n, the length along the surface, we have

$$\frac{\partial\phi}{\partial n} + k\frac{dy}{dn} = 0$$

Now, designating α as the angle that the surface of seepage makes with the x axis and recognizing that

$$\frac{\partial \phi}{\partial n} = u \cos \alpha + v \sin \alpha$$

we obtain

$$u \cos \alpha + v \sin \alpha + k \sin \alpha = 0$$

or

$$v = -u \cot \alpha - k \tag{5}$$

Hence, if the surface of seepage is a straight line ($\alpha =$ constant) in the uv plane, it will be represented as a straight line normal to the surface and passing through the point $(0, -k)$, the lowest point of the circle representing the image of the free surface.

Before proceeding with examples of the construction of hodographs, let us investigate the velocity characteristics in the vicinity of some singular points in the flow domain.

4-2. Flow Characteristics at Singular Points of Flow Domain

Let the z plane in Fig. 4-2a represent a portion of a region of flow at which the boundaries BA and AC intersect at point A. Before assigning

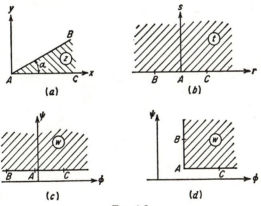

Fig. 4-2

particular aspects to these boundaries, we note that the function

$$t = z^{\pi/\alpha} \tag{1}$$

or

$$z = t^{\alpha/\pi} \tag{2}$$

will provide the correspondence between all points of the z plane and the upper half of the t plane (Fig. 4-2b). Three important cases will be considered.

1. *An angle point at an impervious boundary.* Along an impervious boundary $\psi =$ constant; consequently, BAC of Fig. 4-2a will be a stream-

line and point A will represent an angle point on this streamline. The w plane for this case is shown in Fig. (4-2c). Noting that $w = f(t)$ is everywhere analytic, we can expand this function in the Taylor's series

$$w = c_0 + c_1 t + c_2 t^2 + \cdots \tag{3}$$

where the c's must be complex numbers so that w is complex for all t's. Differentiating Eq. (3) with respect to z, with the help of Eq. (2) we find for the complex velocity

$$W = \frac{dw}{dz} = -\frac{\partial t/\partial z}{\partial t/\partial w} = t^{1-\alpha/\pi}(d_0 + d_1 t + d_2 t^2 + \cdots) \tag{4}$$

where the d's are constants. Thus, at $t = z = 0$, the velocity V along the impervious base is:

$$V = \begin{cases} 0 & \text{if } \alpha < \pi \text{ (Fig. 4-3a)} \\ \infty & \text{if } \alpha > \pi \text{ (Fig. 4-3b)} \\ \text{finite value} & \text{if } \alpha = \pi \text{ (Fig. 4-3c)} \end{cases} \tag{5}$$

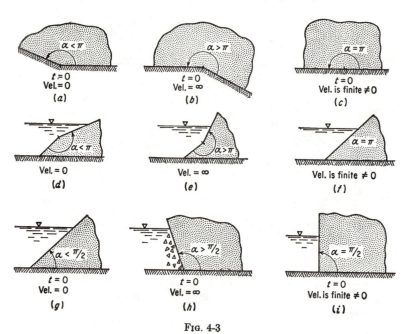

Fig. 4-3

2. *An angle point at a reservoir boundary.* For this case ϕ = constant. Hence in Eq. (3) the c's are real numbers and the results given in (5) are applicable (Fig. 4-3d, e, and f).

3. *Intersection of a reservoir with an impervious boundary.* For this case, AB of Fig. 4-2a will be an equipotential line, AC will be a streamline, and the w plane will be as shown in Fig. 4-2d. Hence the functional relationship between the w and t planes is

$$w = t^{1/2}(e_0 + e_1 t + e_2 t^2 + \cdots) \tag{6}$$

which, making use of Eq. (2), yields for the complex velocity in the vicinity of $t = 0$,

$$W = \frac{dw}{dz} = t^{1/2-\alpha/\pi}(f_0 + f_1 t + f_2 t^2 + \cdots) \tag{7}$$

Thus, at $t = z = 0$, the velocity V at the intersection of the reservoir and the impervious boundary is

$$V = \begin{cases} 0 & \text{if } \alpha < \pi/2 \text{ (Fig. 4-3g)} \\ \infty & \text{if } \alpha > \pi/2 \text{ (Fig. 4-3h)} \\ \text{finite value} & \text{if } \alpha = \pi/2 \text{ (Fig. 4-3i)} \end{cases}$$

4-3. Examples of Velocity Hodographs [116]

As a first example let us consider the upstream section of an earth dam with a horizontal underdrain (Fig. 4-4a). In the hodograph plane (Fig. 4-4b) the region of flow is bounded by the circle $u^2 + v^2 + kv = 0$

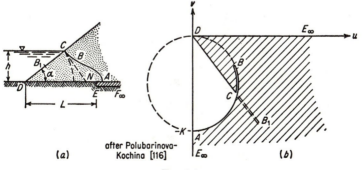

after Polubarinova-
Kochina [116]

(a)   (b)

Fig. 4-4

corresponding to the free surface, a straight line passing through the origin perpendicular to the upstream slope, the u axis corresponding to the impervious base, and the v axis corresponding to the underdrain. The region of flow is shown crosshatched. The slit at point B on the circumference of the circle is a consequence of the velocity at the point of inflection representing the minimum velocity along the free surface (β is a minimum, Fig. 4-1). Generally in earth dams the headwater elevation h is small in comparison to the impervious base L. Therefore, for this condition, the line CN perpendicular to the slope CD at point C would intersect the impervious base of the dam at N (between D and E), and the line of seepage would exhibit a point of inflection.

Let us investigate the movement of the point of inflection B and the resulting hodograph with increasing levels of the headwater h. Assuming the length of the impervious base L to be fixed, we shall investigate the effects of increasing values of the ratio h/L. From Fig. 4-4b it is evident that the velocity at point C represents the maximum velocity along the upstream slope when the free surface exhibits a point of inflection. With increasing values of h/L the point of inflection B will move toward point C, and for some value of h/L the two will coincide and the slit CB of the hodograph will vanish. Computations [113] for two values of the angle α indicate that the point of inflection vanishes at $h/L = 0.820$ for $\alpha = \pi/4$ and $h/L = 0.232$ for $\alpha = \pi/16$. With further increases in h/L the point of maximum velocity B_1 will move down the slope CD, the magnitude of the velocity will increase, and the hodograph will develop the slit CB_1. With additional increases in h/L the slit CB_1 will increase, and as $h/L \to \infty$ the point $B_1 \to \infty$. To show this, we fix h and allow L to approach zero; that is, point E will approach point D (Fig. 4-5). We see at once that the hodograph of Fig. 4-4b for this condition will degenerate into the region AEB_1CA of Fig. 4-5b.

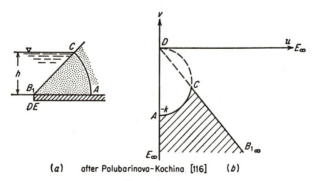

(a)   after Polubarinova-Kochina [116]   (b)

Fig. 4-5

Now let us consider the form of the hodograph when $\alpha = \pi/2$ (Fig. 4-6a). In this case there will be no point of inflection on the free surface and the hodograph will develop as shown in Fig. 4-6b. We note in Fig. 4-6b that the velocity at point D is finite ($\neq 0$), and hence in the hodograph plane the point is displaced along the u axis (Fig 4-3i).

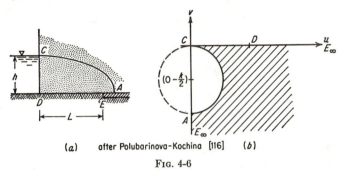

(a)   after Polubarinova-Kochina [116]   (b)

Fig. 4-6

As another example of the construction of the hodograph, let us consider qualitatively the dam of Fig. 4-7a. If θ_0 is less than $\pi/2$ (acute) for small values of h/L, where $h = h_1 - h_2$, a point of inflection B will

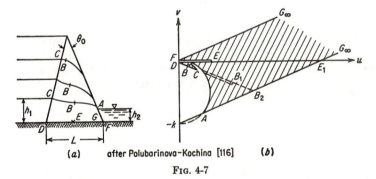

(a)   after Polubarinova-Kochina [116]   (b)

Fig. 4-7

develop on the free surface; the region of flow in the hodograph is shown as the crosshatched portion of Fig. 4-7b. As the ratio h/L increases in this case, the point B will tend toward the point C, then the slit CB_1 will appear, and finally, upon the complete filling of the dam, B_1 will move to B_2, and the section CB_2A of the hodograph will vanish.

If $\theta_0 = \pi/2$, points C and A of the hodograph will coincide and the slit CB_1 will not be possible. Hence a point of inflection will always exist except in the limiting case when the dam is full and the points B and C coincide.

If $\theta_0 > \pi/2$, point C of the hodograph will be below point A (Fig. 4-8), and hence the free surface will always exhibit a point of inflection.

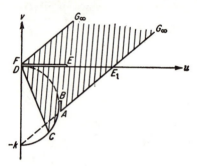

Fig. 4-8

If tail water can be neglected ($h_2 = 0$) in Figs. 4-7 and 4-8, the upper parts of their hodographs will vanish, and point E will move to point E_1.

4-4. Construction of Solution by Complex Velocity

By making use of the complex velocity

$$W = \frac{dw}{dz} = u - iv \tag{1}$$

and the velocity components on the boundaries of the flow region, Polubarinova-Kochina [116] obtained the complete solution for an impervious structure resting on the surface of a flow domain of infinite depth (Fig. 4-9a). A slightly modified form of the solution is given below.

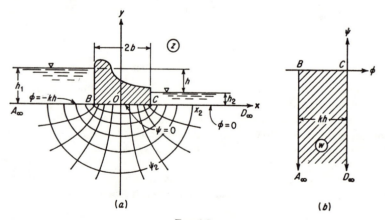

Fig. 4-9

From examination of the requirements of the velocity components along the boundaries of Fig 4-9a, it is evident that along AB and CD the horizontal components of the velocities must be zero ($u = 0$), whereas along the bottom of the structure (BC) the vertical component must vanish ($v = 0$). Hence W is real along BC where $-b < x < b$ and imaginary along AB and CD where $|x| > b$. Considering the functions $(b - z)^{1/2}$ and $(b + z)^{1/2}$, we see that the first function is real for $z = x < b$ and imaginary for $z = x > b$, whereas the second function is real for

$$z = x > -b$$

and imaginary for $z = x < -b$. Therefore, if the two functions are multiplied, we obtain $(b^2 - z^2)^{1/2}$, which yields real values for $-b < x < b$ and imaginary values elsewhere on the x axis ($|x| > b$). Now, assuming that the velocity approaches zero at infinity, the functional relationship

$$W = \frac{dw}{dz} = \frac{M}{\sqrt{b^2 - z^2}} \tag{2}$$

satisfies the required condition of $v = 0$ for $|x| < b$ and $u = 0$ for $|x| > b$. M in Eq. (2) is a real constant, the value of which is to be determined.

Integrating Eq. (2), we obtain

$$w = \phi + i\psi = M \sin^{-1} \frac{z}{b} + N \tag{3}$$

where N is a constant of integration. Considering the correspondence between the w plane (Fig. 4-9b) and the z plane at points C and B, we have the conditions:

At point C: $w = 0$ $z = b$
At point B: $w = -kh$ $z = -b$

These, when substituted into Eq. (3), yield the transformation

$$w = -\frac{kh}{\pi} \cos^{-1} \frac{z}{b} \tag{4}$$

or

$$z = b \cos \frac{\pi w}{kh} \tag{5}$$

To determine the pattern of flow (the flow net), we separate Eq. (5) into the real and imaginary parts

$$x = b \cos \phi' \cosh \psi' \qquad y = -b \sin \phi' \sinh \psi' \tag{6}$$

where $\phi' = \pi\phi/kh$ and $\psi' = \pi\psi/kh$. Now assuming ψ' equal to a sequence of constants, say ψ_n', we find the streamlines to be the ellipses (with foci at the points $\pm b$)

$$\frac{x^2}{b^2 \cosh^2 \psi_n'} + \frac{y^2}{b^2 \sinh^2 \psi_n'} = 1 \tag{7}$$

The equipotential lines will be the confocal hyperbolas

$$\frac{x^2}{b^2 \cos^2 \phi_n'} - \frac{y^2}{b^2 \sin^2 \phi_n'} = 1$$

A portion of the resulting flow net is shown in Fig. 4-9a.

For the velocity distribution under the structure we have, from Eq. (2),

$$u = \frac{kh}{\pi \sqrt{b^2 - x^2}} \tag{8}$$

while along the reservoir boundaries

$$v = \pm \frac{kh}{\pi \sqrt{x^2 - b^2}} \tag{9}$$

where the minus sign applies to AB and the plus sign to CD. The distribution of velocity components is given in Fig. 4-10.

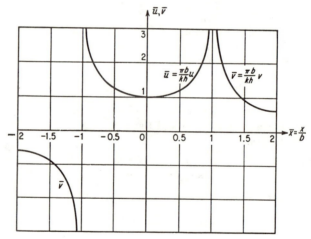

Fig. 4-10. (*After Polubarinova-Kochina* [116].)

From Eq. (9), the exit gradient along CD is found to be

$$i_E = \frac{h}{\pi \sqrt{x^2 - b^2}} \tag{10}$$

which indicates that in the vicinity of $x = b$ the exit gradient is apparently unbounded. Hence there exists in this area the danger of piping. It is for this reason that flat downstream portions of hydraulic structures are set below ground level.

Let us consider now the nature of the uplift pressures acting along the base of the structure. Calling the pressure in excess of tail-water pressure Δp and noting that ϕ was taken as zero along BC (where $y = 0$), from Eq. (4) we have

$$\Delta p = \frac{h\gamma_w}{\pi} \cos^{-1} \frac{x}{b} \tag{11}$$

A plot of $\Delta p/h\gamma_w$ as a function of x/b is given in Fig. 4-11. The excess

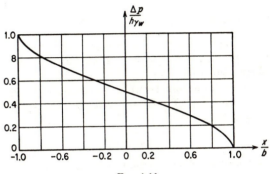

Fig. 4-11

pressure force P (per unit length) acting on the base of the structure is

$$P = \int_{-b}^{b} \Delta p \, dx = \frac{h\gamma_w}{\pi} \int_{-b}^{b} \cos^{-1} \frac{x}{b} \, dx = h\gamma_w b \tag{12}$$

The moment of these uplift forces relative to the heel of the structure (point B) is

$$M_B = \int_{-b}^{b} (x + b) \, \Delta p \, dx = \frac{h\gamma_w}{\pi} \int_{-b}^{b} (x + b) \, \cos^{-1} \frac{x}{b} \, dx = \frac{3h\gamma_w b^2}{4} \tag{13}$$

where the moment arm X of the pressure force (measured from the toe point B) is found to be

$$X = \frac{M_B}{P} = \frac{3b}{4} \tag{14}$$

Example 4-1. In Fig. 4-9, if $h_1 = 20$ ft, $h_2 = 10$ ft, and $b = 30$ ft, find the pressure in the water at point O.

From Fig. 4-11, at $x/b = 0$, $\Delta p/h\gamma_w = 0.5$, and the pressure in the water in excess of tail-water pressure is $\Delta p = 5\gamma_w$. Tail-water pressure (along CD) for this case is $h_2\gamma_w = 10\gamma_w$; hence the pressure in the water at point O is $15\gamma_w$.

Consider now that instead of an infinite depth of porous media we have an impervious boundary coinciding with a streamline (such as ψ_2 in Fig. 4-9a) which terminates at some point x_2 along the tail-water boundary (CD). Recalling that the discharge between any two streamlines is equal to the difference between their stream functions ($q = \Delta\psi$) and taking $\psi = 0$ along the base of the structure, we find from the first of Eqs. (6) with $x = x_2$ (noting that $\phi = 0$ along CD),

$$q = \psi_2 - 0 = \frac{kh}{\pi} \cosh^{-1} \frac{x_2}{b} \tag{15}$$

It is immediately apparent from Eq. (15) that as $x_2 \to \infty$, $q \to \infty$.

4-5. Zhukovsky Functions

A special mapping technique, of particular value when dealing with unconfined flow problems, makes use of an auxiliary transformation called Zhukovsky's function [164].

Noting that the relationship between the velocity potential and the pressure $[\phi = -k(p/\gamma_w + y)]$ can be written as $-kp/\gamma_w = \phi + ky$, if we define $\theta_1 = -kp/\gamma_w$, then

$$\theta_1 = \phi + ky \tag{1}$$

θ_1 is seen to be an harmonic function of x and y as $\nabla^2\theta_1 = \nabla^2\phi \equiv 0$. Hence, its conjugate is the function

$$\theta_2 = \psi - kx \tag{2}$$

Defining $\theta_1 + i\theta_2 = \theta$, we observe that

$$\theta = \theta_1 + i\theta_2 = w - ikz \tag{3}$$

where
$$w = \phi + i\psi \qquad z = x + iy$$

Definition (3) and any function with its real or imaginary part differing from it by a constant multiplier is called a *Zhukovsky function*. Nelson-Skornyakov, whose work will be considered in a later chapter, uses the modified form of Zhukovsky's function

$$G = z - \frac{iw}{k} \tag{4}$$

To illustrate the potential usefulness of Zhukovsky's function, consider the flow around an impervious sheetpile founded in an infinite depth of porous media (Fig. 4-12a). Here, instead of specifying confinement, the boundary DE_∞ will be allowed to develop as a free surface. The corresponding w plane is given in Fig. 4-12b. The free surface must satisfy the conditions $\phi + ky = 0$ and $\psi = 0$. Consequently, Zhukovsky's function for the free surface will have $\theta_1 = 0$ and $\theta_2 = -kx$, and hence in the θ plane the free surface will have as its image the negative imaginary axis (Fig. 4-12c). Along the sheetpile BCD, $\psi = 0$ and $x = 0$; hence $\theta_2 = 0$; its image is along the negative real axis in the θ plane. Finally, along AB, where $y = 0$, $\phi = -kh_1$, and $-\infty \leqq \psi \leqq 0$, we have $\theta_1 = -kh_1$

and $-\infty \leqq \theta_2 \leqq 0$; its image is the line BA_∞ in Fig. 4-12c. Thus, we see that the region of flow (the z plane) with an unknown boundary (DE) is transformed into an infinite strip in Zhukovsky's θ plane.

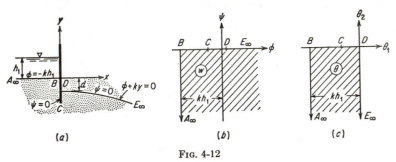

Fig. 4-12

If, now, the functional relationship between the θ plane and the w plane were determined, say $w = f(\theta)$, then, making use of Eq. (3), θ could be eliminated, and we would have $w = f(z)$, from which, as in Sec. 4-4, all pertinent seepage characteristics could be found. The method for transforming one polygon conformally onto another is provided by the Schwarz-Christoffel transformation, which will now be developed. The solution to the problem given above will be completed in Chap. 7.

4-6. The Schwarz-Christoffel Transformation

In Chap. 3 we considered the mapping of various simple shapes where the transformations were known a priori. In groundwater problems, where it is often necessary to determine the seepage characteristics within complicated boundaries, this procedure is far from satisfactory. Theoretically, the transformation exists which will map any pair of simply connected regions conformally onto each other. This is assured by the Riemann mapping theorem [101]; however, the determination of a general solution for the mapping problem has thus far defied discovery. At first this may appear somewhat disturbing; however, as in the case of the Zhukovsky functions, the use of appropriate auxiliary mapping techniques enables us to transform even complicated flow regions into regular geometric shapes. Generally these figures will be polygons having a finite number of vertices (one or more of which may be at infinity). Thus the method of mapping a polygon from one or more planes onto the upper half of another plane is of particular importance.

If a polygon is located in the z plane, then the transformation that maps it conformally onto the upper half of the t plane ($t = r + is$) is

$$z = M \int \frac{dt}{(t - a)^{1-A/\pi}(t - b)^{1-B/\pi}(t - c)^{1-C/\pi} \cdots} + N \qquad (1)$$

where M and N are complex constants, A, B, C, . . . , are the interior angles (in radians) of the polygon in the z plane (Fig. 4-13a), and a, b, c, . . . ($a < b < c < \cdots$) are points on the real axis of the t plane corresponding to the respective vertices A, B, C, . . . (Fig. 4-13b). We

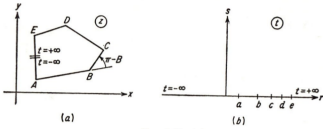

(a)   (b)

Fig. 4-13

note, in particular, that the complex constant N corresponds to the point on the perimeter of the polygon that has its image at $t = 0$. Equation (1) is called the *Schwarz-Christoffel transformation* in honor of the two mathematicians, the German H. A. Schwarz (1843–1921) and the Swiss E. B. Christoffel (1829–1900), who discovered it independently.

The transformation can be considered as the mapping of a polygon from the z plane onto a *similar* polygon in the t plane in such a manner that the sides of the polygon in the z plane extend through the real axis of the t plane. This is accomplished by opening the polygon at some convenient point, say between A and E of Fig. 4-13a, and extending one side to $t = -\infty$ and the other to $t = +\infty$ (Fig. 4-13b). In this operation the sides of the polygon are bent into a straight line extending from $t = -\infty$ to $t = +\infty$ and are placed along the real axis of the t plane. The interior angle at the point of opening may be regarded as π (in the z plane) and, as noted in Eq. (1), takes no part in the transformation. The point of opening in the z plane is represented in the upper half of the t plane by a semicircle with a radius of infinity. Thus the Schwarz-Christoffel transformation, in effect, maps conformally the region interior to the polygon $ABC \cdots$ of the z plane into the interior of the polygon bounded by the sides ab, bc, . . . and a semicircle with a radius of infinity in the upper half of the t plane, or, more simply, into the entire upper half of the t plane.

To demonstrate the mechanism of the Schwarz-Christoffel transformation, we recall that in Sec. 3-3 it was shown that a derivative of the form dz/dt could be considered as a complex operator that transforms an element of t, by rotation and magnification, into a corresponding element in z. Thus, writing Eq. (1) as

$$\frac{dz}{dt} = M(t - a)^{A/\pi-1}(t - b)^{B/\pi-1}(t - c)^{C/\pi-1} \cdots \tag{2}$$

and taking the arguments of both sides, we find that any section of the real axis of t (where arg $dt = 0$) will be rotated in z by

$$\text{arg } dz = L + \left(\frac{A}{\pi} - 1\right) \text{arg } (t - a) + \left(\frac{B}{\pi} - 1\right) \text{arg } (t - b)$$
$$+ \left(\frac{C}{\pi} - 1\right) \text{arg } (t - c) + \cdots \quad (3)$$

where $L = \text{arg } M$ is a constant. Let us now consider the rotation of elements along the real axis of t, excluding the terminal points $a, b, c, \ldots$. For $a < t < b$, arg $(t - a) = 0$ since $(t - a)$ is real and positive, and arg $(t - b) = \text{arg } (t - c) = \cdots = \pi$ since $(t - b), (t - c), \ldots,$ are all real and negative. Thus, for $a < t < b$, arg dz is a constant equal to

$$\text{arg } dz = L + B - \pi + C - \pi + \cdots$$

which shows that this section of the t plane has its image in the z plane along a straight line (AB in Fig. 4-13a). In like manner, for $b < t < c$, we have

$$\text{arg } dz = L + C - \pi + \cdots$$

which will have its image along the straight line BC; that is, arg dz for $b < t < c$ exceeds arg dz for $a < t < b$ by the positive angle $\pi - B$, which is precisely the deflection angle at point B. As t moves from $-\infty$ to $+\infty$ along the real axis, it is seen that z completes its circuit through the total external angular change of 2π radians and hence encloses the polygon $ABC \cdots$. The complex constants M and N of Eq. (1) merely control the size and position of the polygon.

The validity of the transformation at the points $t = a$, $t = b$, $\ldots$ remains to be investigated. As t moves along the real axis through the point b, $(t - b)$ changes from a negative to a positive number and arg $(t - b)$ decreases from π to 0. Hence the third term in Eq. (3) changes by $(B/\pi - 1)(-\pi) = \pi - B$, which, as was noted previously, is the value of the positive deflection angle at the vertex B (Fig. 4-14a). In effect,

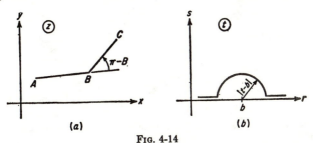

Fig. 4-14

as z passes through the vertex B, its image in the t plane passes around an indented semicircle at point b (Fig. 4-14b), the radius of which ($|t - b|$) can be made as small as we wish by adjusting $|M|$.

In many problems we shall place one or more vertices of the polygon in the z plane at infinity in the t plane. If, for example, $a \to -\infty$, we can take the complex constant M to be of argument L and modulus $C(-a)^{-A/\pi+1}$, so that Eq. (2) becomes*

$$\frac{dz}{dt} = Ce^{iL}(-a)^{1-A/\pi}(t-a)^{A/\pi-1}(t-b)^{B/\pi-1} \cdots$$

or

$$\frac{dz}{dt} = Ce^{iL}\left(\frac{t-a}{-a}\right)^{A/\pi-1}(t-b)^{B/\pi-1} \cdots$$

Now, as $a \to -\infty$, $[(t-a)/-a]^{A/\pi-1} \to 1$, and hence we see that factors corresponding to vertices at infinity in the t plane do not appear in the transformation.

On the basis of the foregoing it follows that corresponding values of a, b, c, . . . and A, B, C, . . . can be chosen so that the polygons in their respective planes are similar. It can be shown [101] that any three of the values a, b, c, . . . can be chosen arbitrarily to correspond to three of the vertices of the given polygon A, B, C, ... The $(n-3)$ remaining values must then be determined so as to satisfy conditions of similarity. Whereas we shall often choose to map a vertex of the flow region (z plane) into one at infinity in the t plane, it is important to note that not only is this factor omitted from the transformation, but the number of arbitrary values is reduced by 1.

4-7. Examples of Schwarz-Christoffel Mappings

Consider the mapping of the semi-infinite strip $A_\infty BCA_\infty$, of width h, onto the upper half of the t plane (Fig. 4-15). The strip may be considered as a triangle with interior angles $B = \pi/2$, $C = \pi/2$, and $A = 0$.

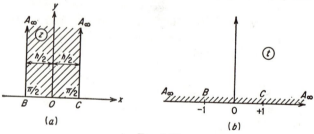

Fig. 4-15

In mapping the points A_∞, B, and C arbitrarily on the points $t = -\infty$, $t = -1$, and $t = +1$, respectively, conditions of symmetry will place the fourth vertex at $t = +\infty$, and the Schwarz-Christoffel transformation will be

$$z = M \int \frac{dt}{(t+1)^{1-\frac{1}{2}}(t-1)^{1-\frac{1}{2}}} + N \tag{1}$$

* This definition does not restrict M; it simply defines C.

We note immediately in this expression that as $z = t = 0$, $N = 0$; hence

$$z = M \int \frac{dt}{\sqrt{t^2 - 1}} = -iM \int \frac{dt}{\sqrt{1 - t^2}} = -iM \sin^{-1} t$$

To determine the constant M, we set the corresponding conditions, $t = 1$ for $z = h/2$, which yield $M = ih/\pi$, and hence the required transformation is

$$z = \frac{h}{\pi} \sin^{-1} t \tag{2}$$

or

$$t = \sin \frac{\pi z}{h} \tag{3}$$

Consider now the mapping of the semi-infinite strip $A_\infty BCA_\infty$ of Fig. 4-16a onto the upper half of the t plane.   The Schwarz-Christoffel trans-

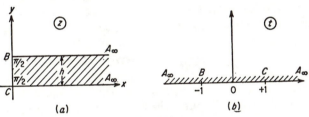

FIG. 4-16

formation for this case is the same as Eq. (1).   However, it is convenient to take the integral as

$$z = M \int \frac{dt}{\sqrt{t^2 - 1}} + N = M \cosh^{-1} t + N$$

At points C, where $z = 0$ corresponds to $t = 1$, we have

$$0 = M \cosh^{-1} 1 + N \qquad \text{or} \qquad N = 0$$

Also, at points B, $z = ih$ corresponds to $t = -1$, and

$$ih = M \cosh^{-1} (-1)$$

Noting that, $\cosh^{-1} (-1) = i\pi$ [Eq. (22a) Appendix A], we find

$$ih = Mi\pi \qquad \text{and} \qquad M = \frac{h}{\pi}$$

Therefore the required transformation is

$$z = \frac{h}{\pi} \cosh^{-1} t \tag{4}$$

or

$$t = \cosh \frac{z\pi}{h} \tag{5}$$

Finally, we shall consider the mapping of the region shown in Fig. 4-17a, with a slit of height h, onto the upper half plane of t (Fig. 4-17b).

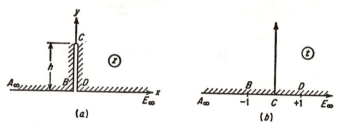

Fig. 4-17

Let us map arbitrarily the three points A_∞, B, and C onto the points $t = -\infty$, $t = -1$, and $t = 0$, respectively. Conditions of symmetry will place the points D and E_∞ on $t = +1$ and $t = +\infty$. The internal angles are $\pi/2$ at B and D, and 2π at C; hence

$$z = M \int \frac{dt}{(t+1)^{1/2} t^{-1} (t-1)^{1/2}} + N = M\sqrt{t^2 - 1} + N$$

Now at points D, since $z = 0$ and $t = 1$, $N = 0$. At points C, $z = ih$ and $t = 0$; hence $M = h$. Therefore the required transformation is

$$z = h\sqrt{t^2 - 1} \qquad (6)$$

or

$$t = \pm \frac{1}{h}\sqrt{z^2 + h^2} \qquad (7)$$

PROBLEMS

1. For the transformation $w = 2z - z^2$, (a) determine the magnitude and direction of the resultant velocity at the point $z = 1 + 2i$; (b) find where stagnation will occur; (c) find the locus of points where the resultant velocity makes an angle of 45° with the horizontal.

2. For each of the following transformations, determine the x and y coordinates of the points of stagnation and/or turbulence.

(a) $w = 4z^2 + 6z + 5$ (b) $w = \ln z$

(c) $w = \cos z$ (d) $w = \tan^{-1} z$

(e) $w = \sinh^{-1} z$ (f) $w = \cosh^{-1} z$

3. Under the transformation $w = z^2$, determine the equation of the curve along which the resultant velocity equals 9.

4. Discuss the influence on point D of the hodograph of Fig. 4-6 for (a) $h/L \rightarrow \infty$; (b) $h/L \rightarrow 0$.

5. Demonstrate on the basis of the velocity hodograph that, in practice, an homogeneous dam on an impervious base without upstream slope protection and with a slope drain will always exhibit a point of inflection on the free surface.

6. Obtain the velocity hodographs for each of the sections in Fig. 4-18.

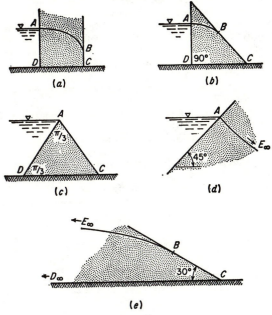

Fig. 4-18

7. Find the functional relationship between the z plane and the hodograph plane in Fig. 4-18c if the interior angles at C and D are both $45°$.

8. Obtain the velocity hodograph for the sections in Fig. 4-19 and determine the magnitude of the velocity at the points A, C, and E. $k = 0.006$ cm/sec.

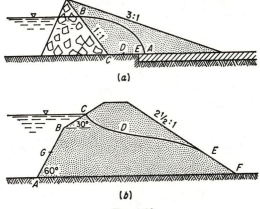

Fig. 4-19

9. In Fig. 4-9a, $2b = 20$ ft, $h_1 = 8$ ft, and $h_2 = 0$. The section contains an impervious lower boundary in the shape of an ellipse with foci at $\pm b$ and with a major axis of 40 ft. Determine (a) the discharge if $k = 0.002$ cm/sec; (b) the velocity (magnitude and direction) at the point $z = 15 - 10i$; (c) the pressure force on the bottom of the structure; (d) the factor of safety with respect to piping at $z = x = 15$ ft.

10. On the basis of Sec. 4-4, demonstrate that whatever form a structure founded on the surface of a porous layer of infinite depth may have, at an infinite depth the boundary flow line will be circular.

11. Obtain the image of Fig. 4-12a under Nelson-Skornyakov's form of Zhukovsky's function [Eq. (4), Sec. 4-5].

12. Where is the region exterior to a polygon in the z plane mapped in the t plane by the Schwarz-Christoffel transformation?

13. Explain why the Schwarz-Christoffel transformation is said to map a polygon from the z plane onto Im $(z) > 0$ and not Im $(z) \geqq 0$.

14. Obtain the transformation of the infinite strip in Fig. 4-16 if point C is not at the origin but at the point $z = 1 + 2i$.

15. Verify Eq. (3), Sec. 4-7, for point A in Fig. 4-15a.

16. Verify each of the Schwarz-Christoffel transformations in Fig. 4-20.

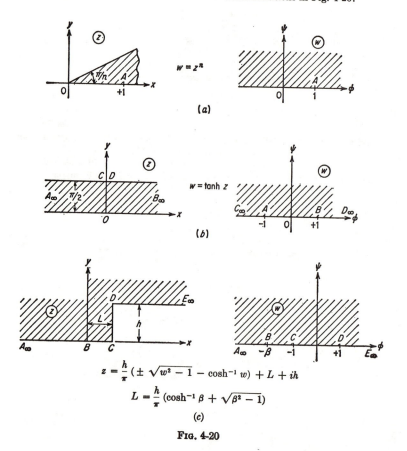

$$z = \frac{h}{\pi}(\pm \sqrt{w^2 - 1} - \cosh^{-1} w) + L + ih$$

$$L = \frac{h}{\pi}(\cosh^{-1} \beta + \sqrt{\beta^2 - 1})$$

(c)

Fɪɢ. 4-20

5

Confined Flow

5-1. General Discussion

Having investigated the fundamentals of groundwater flow and the main properties and techniques of conformal mapping, we shall, in the following chapters, make use of these to derive the solutions of some important classes of groundwater flow problems. The class of problems to be investigated in this chapter are characterized in that all boundaries of the flow domain are completely defined. As noted previously in Sec. 2-1, such flow is said to be *confined*.

In the preceding chapters we saw that all desired flow characteristics could be obtained once the function $w = f(z)$ was known. A general method of determining this functional relationship for confined flow problems, including the closed-form solutions of many such problems, was first published in Russian by Pavlovsky [110] in 1922. However, because of language difficulties, this work has remained relatively unknown to the English-speaking engineer. To date, only a brief digest of Pavlovsky's work has appeared in English [86] and almost nothing is available of the more recent developments of his disciples. It is somewhat disconcerting to note in this literature solutions to problems that have since been and still are being attempted by others.

To demonstrate Pavlovsky's general method of solution for $w = f(z)$, consider the structure (dam, weir, or spillway) shown in Fig. 5-1a (z plane). The boundaries of the flow region are of two types: (1) impervious boundaries (ψ = constant), and (2) reservoir boundaries (ϕ = constant). The corresponding w plane, bounding the flow region, is the rectangle $ABCD$ in Fig. 5-1b. As the z plane and w plane are both polygons (vertices A and D of the z plane are at infinity), by means of the Schwarz-Christoffel transformation the flow region in each of these planes can be mapped conformally onto the same half of an auxiliary t plane (Fig. 5-1c), thereby yielding, say, the functions $z = f_1(t)$ and $w = f_2(t)$. Then, either by eliminating the variable t or by using t as a parameter, the function $w = f(z)$ can be established.

The foregoing method is particularly well suited for the analytical determination of the exit gradient. To illustrate, we note that the gradient at any point in an isotropic flow region is

$$I = -\frac{dh}{ds} = \frac{1}{k}\frac{d\phi}{ds} = \frac{1}{k}\frac{d\phi}{dt}\frac{dt}{dz}\frac{dz}{ds} \qquad (1)*$$

where s is the direction of the streamline at that point. Defining the angle between the direction of the streamline and the x axis as θ, we have $dz/ds = \cos\theta + i\sin\theta$. Since the streamline at the critical exit point (such as point C of Fig. 5-1a) generally represents $\psi = $ constant (hence

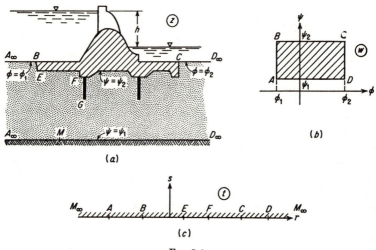

Fig. 5-1

$d\phi/dt = dw/dt$) and intersects the tail-water equipotential boundary at 90° ($\theta = 90°$), Eq. (1) will generally reduce to

$$I_E = \frac{i}{k}\left(\frac{dw}{dt}\frac{dt}{dz}\right)_{\psi=\text{constant}} \qquad (2)†$$

Recognizing the derivatives in Eq. (2) as statements of the Schwarz-Christoffel transformation [Eq. (2), Sec. 4-6], we see that the exit gradient is all but known once the constants M of the transformations have been determined.

---

* I will be used to denote a gradient and I_E the exit gradient rather than the customary i [Eq. (2), Sec. 1-4] to prevent confusion with the imaginary $i = \sqrt{-1}$.

† To account for the deviations and *uncertainties in nature*, Khosla, Bose, and Taylor [69] recommend that the following factors of safety be applied as critical values of exit gradients: gravel, 4 to 5; coarse sand, 5 to 6; fine sand, 6 to 7.

5-2. Hydraulic Structure on Surface of Infinite Depth of Porous Media*

As a first example of the solution of confined flow problems we shall consider a structure with a single sheetpile s resting on the surface of an infinite depth of porous material. To provide the most general solution to this problem, we shall choose the z plane as shown in Fig. 5-2a. In

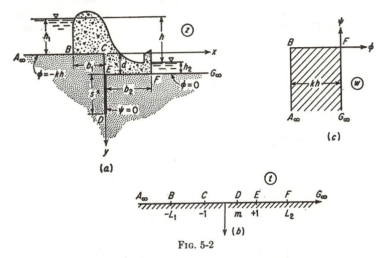

FIG. 5-2

this figure, $BCDEF$ represents the bottom contour of the structure, including the sheetpile CDE. Once the general solution is determined, various special cases can be obtained by adjusting the lengths b_1, b_2, d, and s. For example, taking $d = s = 0$, we have as a special case the problem solved previously in Sec. 4-4 (Fig. 4-9a).

We begin by mapping the z plane (Fig. 5-2a) onto the lower half of the t plane (Fig. 5-2b). For simplicity we shall map the points A_∞, C, and E onto the points $t = -\infty$, -1, $+1$ of the real axis of the t plane. Considerations of symmetry will place G_∞ at $t = +\infty$. The points B, F, and D will be located on the points $t = -L_1$, L_2, and m, which are to be determined. Hence we have for the Schwarz-Christoffel transformation

$$z = M \int \frac{(t - m)\, dt}{\sqrt{t^2 - 1}} + N = M\sqrt{t^2 - 1} - Mm \cosh^{-1} t + N \quad (1)\dagger$$

To evaluate the constants N, m, and M, we consider the corresponding values of z and t at the vertices E, C, and D. For E, $z = id$ and $t = +1$;

---

* Although the solution given here follows more closely that of Khosla, Bose, and Taylor [69], it is interesting to note that Pavlovsky had solved the same problem, but in much less detail, 14 years earlier.

† Note that $\cosh^{-1} t = \ln (t + \sqrt{t^2 - 1})$.

therefore $N = id$. For C, $z = 0$, $t = -1$, and $Mm = d/\pi$. At points D, where $z = i(d + s)$ and $t = m \leqq 1$, we get, substituting for N and M,

$$\frac{s\pi}{d} = \frac{\sqrt{1 - m^2}}{m} - \cos^{-1} m \tag{2}$$

Equation (2) allows for the determination of the parameter m as a function of s/d, and hence, in any problem, m may be considered as a known quantity. To facilitate computations, a plot of this relation is given in Fig. 5-3.

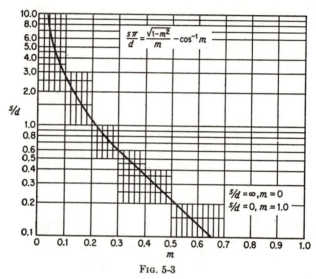

FIG. 5-3

Substituting for M and N into Eq. (1), we obtain for the mapping of the z plane onto the t plane,

$$z = \pm \frac{d}{\pi m} \sqrt{t^2 - 1} - \frac{d}{\pi} \cosh^{-1} t + id \tag{3}$$

Now for the mapping of the w plane (Fig. 5-2c) onto the lower half of the t plane, we have from the Schwarz-Christoffel transformation (cf. Eq. 380.001, Ref. 29),

$$\begin{aligned} w &= M_1 \int \frac{dt}{(t + L_1)^{1/2}(t - L_2)^{1/2}} + N_1 \\ &= iM_1 \sin^{-1} \frac{2t + L_1 - L_2}{L_1 + L_2} + N_1 \\ &= M_2 \sin^{-1} \frac{t + \lambda_1}{\lambda} + N_1 \end{aligned} \tag{4}$$

where $\lambda_1 = (L_1 - L_2)/2$ and $\lambda = (L_1 + L_2)/2$. At points F, $t = L_2$ and $w = 0$; therefore $N_1 = -M_2 \pi/2$. For B, $t = -L_1$ and $w = -kh$;

hence $M_2 = kh/\pi$ and

$$t = \lambda \cos \frac{w\pi}{kh} - \lambda_1 \qquad (5)$$

To determine L_1 and L_2 and hence λ and λ_1, we note that Eq. (3) provides the correspondence between all points in the z and t planes. Thus, taking $z = -b_1$ and $t = -L_1$ (points B) and calling $L_1 = \cosh \beta_1$, after some elaboration we find

$$\frac{-b_1\pi m}{d} = \sinh \beta_1 + m\beta_1 \qquad (6)$$

where m is given by Eq. (2) or Fig. 5-3. By a similar procedure, from the correspondence at points F, with $L_2 = \cosh \beta_2$, we get

$$\frac{b_2\pi m}{d} = \sinh \beta_2 - m\beta_2 \qquad (7)$$

Because of the implicit nature of these equations a graphical method of solution is recommended. The method of solution will be illustrated below.

Finally, substituting Eq. (5) for t into Eq. (3), we get for the required transformation between all points in the z and w planes,

$$z = \frac{d}{\pi m}\left[\left(\lambda \cos \frac{w\pi}{kh} - \lambda_1\right)^2 - 1\right]^{1/2} - \frac{d}{\pi}\cosh^{-1}\left(\lambda \cos \frac{w\pi}{kh} - \lambda_1\right) + id \qquad (8)$$

In particular, noting that $\psi = 0$ along the contour of the structure (Fig. 5-2a), the correspondence between any point along this boundary and the potential at that point is

$$z = \frac{d}{\pi m}\left[\left(\lambda \cos \frac{\phi\pi}{kh} - \lambda_1\right)^2 - 1\right]^{1/2} - \frac{d}{\pi}\cosh^{-1}\left(\lambda \cos \frac{\phi\pi}{kh} - \lambda_1\right) + id \qquad (9)$$

For design purposes we need to know the pressure distribution in the water acting along the various sections of the structure and the magnitude of the exit gradient. We consider first the question of pressure distribution. As the y axis was taken positive down (Fig. 5-2a),

$$\phi = -k\left(\frac{p}{\gamma_w} - y\right) + C$$

Now, noting that at $y = 0$, $\phi = -kh$ and $p/\gamma_w = h_1$, we find

$$C = -k(h - h_1)$$

Whereas, from Eq. (5), along the impervious boundary ($\psi = 0$),

$$\phi = -\frac{kh}{\pi}\cos^{-1}\frac{t + \lambda_1}{\lambda} \qquad 0 \leqq \cos^{-1}\frac{t + \lambda_1}{\lambda} \leqq \pi$$

the pressure at any point along this boundary will be given by

$$p = \gamma_w\left(\frac{h}{\pi}\cos^{-1}\frac{t + \lambda_1}{\lambda} + y - h + h_1\right) \qquad (10)$$

In particular, at the points C, D, and E, respectively, we find

$$p_C = \gamma_w \left(\frac{h}{\pi} \cos^{-1} \frac{\lambda_1 - 1}{\lambda} - h + h_1 \right) \tag{11a}$$

$$p_D = \gamma_w \left(\frac{h}{\pi} \cos^{-1} \frac{\lambda_1 + m}{\lambda} + s + h_2 \right) \tag{11b}$$

$$p_E = \gamma_w \left(\frac{h}{\pi} \cos^{-1} \frac{\lambda_1 + 1}{\lambda} + h_2 \right) \tag{11c}$$

Example 5-1. Determine the pressure in the water at point C of the structure in Fig. 5-4a.

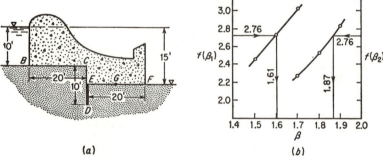

(a)   (b)

Fig. 5-4

From the given data we have $s/d = 1$; hence from Fig. 5-3, $m = 0.22$. Thus $-b_1\pi m/d = -2.76$ and $b_2\pi m/d = 2.76$. Now, taking the left sides of Eqs. (6) and (7) as

$$f(\beta_1) = \sinh \beta_1 + m\beta_1$$
$$f(\beta_2) = \sinh \beta_2 - m\beta_2$$

and assuming a sequence of β's, we construct Table 5-1.

Table 5-1

β	$\sinh \beta$	$m\beta$	$f(\beta_1)$	$f(\beta_2)$
1.5	2.13	0.33	2.46	
1.6	2.37	0.35	2.72	
1.7	2.65	0.37	3.02	2.28
1.8	2.94	0.40	...	2.54
1.9	3.27	0.42	...	2.85

Plotting the values of $f(\beta)$ as a function of β (Fig. 5-4b), we find $\beta_1 = 1.61$ and $\beta_2 = 1.87$. Hence, $L_1 = \cosh \beta_1 = 2.60$, $L_2 = \cosh \beta_2 = 3.32$, $\lambda = 2.96$, and $\lambda_1 = -0.36$. Thus, from Eq. (11a), the pressure at point C is

$$p_C = \gamma_w \left[\frac{15}{\pi} \cos^{-1}(-0.46) - 15 + 10 \right] = 4.8\gamma_w$$

Let us consider now the particular sections shown in Fig. 5-5* (where $d = 0$). From Eq. (2) we see that for this case $m = 0$ also. Hence, in

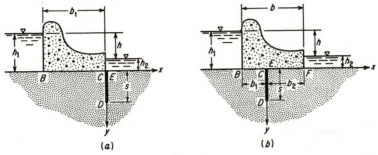

Fig. 5-5

Eq. (3), we have an indeterminate form for the first term in the R.H.S.

$$z = \frac{d}{\pi m} \sqrt{t^2 - 1}$$

To resolve this indeterminacy, we substitute for d/m from Eq. (2), obtaining

$$z = \frac{s \sqrt{t^2 - 1}}{\sqrt{1 - m^2} - m \cos^{-1} m}$$

which, letting $m \to 0$, yields the transformation between the z plane and t plane:

$$z = s \sqrt{t^2 - 1} \tag{12}†$$

Now, noting in Fig. 5-5a that at points B, $z = -b_1$ and $t = -L_1$, and at points F that $z = b_2 = 0$ and $t = L_2$, we get $L_1 = [1 + (b_1/s)^2]^{1/2}$ and $L_2 = 1$; hence

$$\lambda = \frac{1}{2}\left[1 + \sqrt{1 + \left(\frac{b_1}{s}\right)^2}\right] \tag{13a}$$

$$\lambda_1 = \frac{1}{2}\left[\sqrt{1 + \left(\frac{b_1}{s}\right)^2} - 1\right] \tag{13b}$$

For the section shown in Fig. 5-5b it is easy to verify that

$$\lambda = \frac{1}{2}\left[\sqrt{1 + \left(\frac{b_1}{s}\right)^2} + \sqrt{1 + \left(\frac{b_2}{s}\right)^2}\right] \tag{14a}$$

$$\lambda_1 = \frac{1}{2}\left[\sqrt{1 + \left(\frac{b_1}{s}\right)^2} - \sqrt{1 + \left(\frac{b_2}{s}\right)^2}\right] \tag{14b}$$

* These sections were also investigated in some detail by Weaver [160] in 1932.
† Compare this equation with Eq. (6), Sec. 4-7.

Substituting for t from Eq. (12) into Eq. (10), we find that the pressure along the base of the sections ($y = 0$) in Fig. 5-5 is given by

$$p = \gamma_w \left(\frac{h}{\pi} \cos^{-1} \frac{\lambda_1 s \pm \sqrt{s^2 + x^2}}{\lambda s} + h_2 \right) \quad (15a)$$

and along the sheetpile ($x = 0$) by

$$p = \gamma_w \left(\frac{h}{\pi} \cos^{-1} \frac{\lambda_1 s \pm \sqrt{s^2 - y^2}}{\lambda s} + y + h_2 \right) \quad (15b)$$

In Fig. 5-6 are shown curves prepared by Khosla, Bose, and Taylor [69] that give the pressure in the water Δp in excess of hydrostatic pressure relative to tail-water elevation at the points C, D, and E of Fig. 5-5b

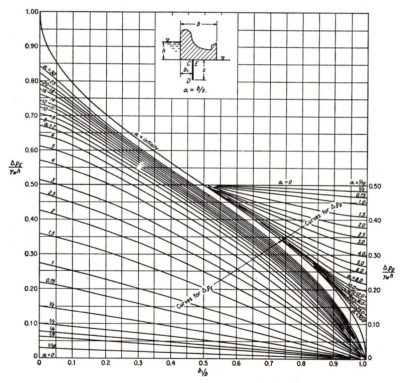

Fig. 5-6. (*After Khosla, Bose, and Taylor* [69].)

for any position of the sheetpile. The values of Δp at point E (Δp_E) can be obtained directly from the plot. For example, for a sheetpile of $s = b/2$ ($\alpha = b/s = 2$) located one-fourth of the base length from the heel of the structure ($b_1/b = \frac{1}{4}$), the pressure in the water at point E,

in excess of tail-water hydrostatic pressure, will be $\Delta p_E = 0.35\ h\gamma_w$. In like manner, pressures at point D for $b_1/b \geqq \frac{1}{2}$ can be obtained directly from the curves labeled "curves for Δp_D" on the figure. For values of $b_1/b < \frac{1}{2}$, the abscissa is entered with values of $1 - b_1/b$, and the desired value of $\Delta p_D/\gamma_w h$ is obtained by subtracting the determined value from unity. Thus, for the example given above, $\Delta p_D = 0.57\ h\gamma_w$. To find the excess pressure at point C, the abscissa is entered with $1 - b_1/b$, and the value of $\Delta p_E/h\gamma_w$ is determined for the proper α from the Δp_E curves. Then the desired $\Delta p_C/h\gamma_w$ is obtained by subtracting this value from unity. For the example above, $\Delta p_C = 0.87\ h\gamma_w$.

Weaver [160] obtained plots of the total uplift force and the total uplift moment (with respect to the heel) for various positions of the sheetpile, as shown in Figs. 5-7a and b, respectively. In these figures both the total uplift force and the moment are expressed as per cent of the force and moment with the sheetpile absent and without tail water [Eqs. (12) and (13), Sec. 4-4]. We note from these figures that both the force and moment decrease as the piling is positioned close to the heel. Also, for sheetpiles set at the center of the base, the total uplift force is independent of the depth of embedment. A similar region occurs for the total moment at a position slightly downstream of the center.

Turning now to the determination of the exit gradient and recalling from Sec. 4-4 that the exit gradient is unbounded at the toe of a structure set at the ground level, we need to consider only those sections with $b_2 = 0$, as shown in Fig. 5-8. From Eq. (2), Sec. 5-1, the exit gradient

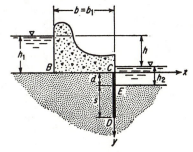

Fig. 5-7. $\alpha = b/s$ (see Fig. 5-5b). (*After Weaver* [160].)

Fig. 5-8

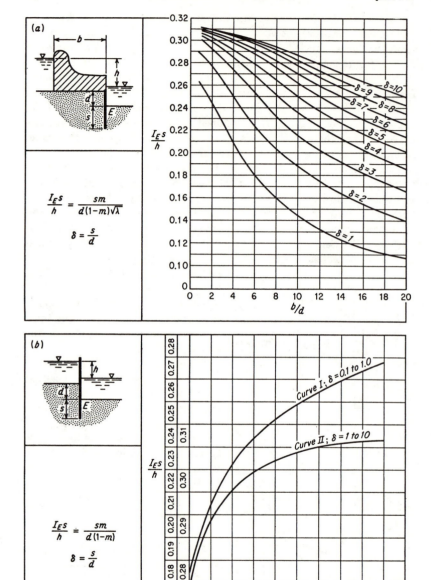

FIG. 5-9. (*After Khosla, Bose, and Taylor* [69].)

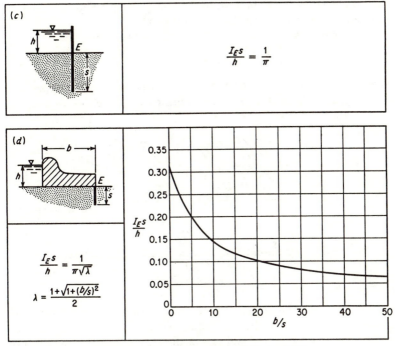

FIG. 5-9. (*After Khosla, Bose, and Taylor* [69].)

(point E) for this case is

$$I_E = \frac{i}{k} \frac{dw}{dt} \frac{dt}{dz}$$

Now, noting that $dw/dt = M_1/[(t + L_1)^{1/2}(t - L_2)^{1/2}]$ and $M_1 = kh/i\pi$ [Eq. (4)], $dz/dt = M(t - m)/(t^2 - 1)^{1/2}$ and $M = d/m\pi$ [Eq. (1)], and for this case $L_2 = 1$, we have for the exit gradient at point E (where $t = 1$),

$$I_E = \frac{mh}{d(1 - m)\sqrt{\lambda}} \tag{16}$$

Curves giving the variation of the exit gradient, for different combinations of structural parameters, are shown in Fig. 5-9.

5-3. Inclined Sheetpile

We shall next consider the problem of flow around an inclined sheetpile founded in a porous base of infinite depth. This problem is of interest not only for the determination of the flow around oblique cutoffs but also for investigation of vertical sheetpiles in anisotropic soils. The solution to the problem was first obtained by Verigin [155] by mapping a domain with radial slits onto an auxiliary upper half plane.

The mapping was accomplished by using the transformation

$$z = Ce^{i\pi\alpha_n}(a_1 - t)^{\alpha_n - \alpha_1}(a_2 - t)^{\alpha_1 - \alpha_2} \cdots (a_k - t)^{\alpha_{n-1} - \alpha_n} \tag{1}$$

which maps a region with radial slits in the z plane onto the upper half of the t plane (Fig. 5-10). In this expression C is a real constant, $\pi\alpha_1$,

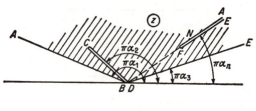

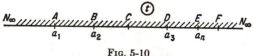

Fig. 5-10

$\pi\alpha_2$, $\pi\alpha_3$, . . . are the angles that the sides of the slits make with the abscissa, $\pi\alpha_n$ is the particular angle of the slit $ANFE$, $a_1, a_2, a_3, \ldots, a_n$ are the image points on the real axis of the t plane, and the point N has its image at $t = \infty$.

To demonstrate the validity of the mapping: taking the logarithm of Eq. (1), we obtain

$$\ln z = \ln |z| + i \arg z = \ln C + \pi\alpha_n i + (\alpha_n - \alpha_1) \ln (a_1 - t)$$
$$+ (\alpha_1 - \alpha_2) \ln (a_2 - t) + \cdots + (\alpha_{n-1} - \alpha_n) \ln (a_n - t)$$

Now, considering real values of t between $-\infty$ and a_1, we see that

$$\arg (a_1 - t) = \cdots = \arg (a_n - t) = 0$$

and hence the imaginary part of Eq. (1) reduces to $\arg z = \pi\alpha_n$, which corresponds to the direction of the line NA (Fig. 5-10) in the z plane. Likewise, selecting a real value of t between a_1 and a_2, we obtain $\arg w = \pi\alpha_1$, which corresponds to the direction of the line AB. Repeating this procedure for other sections on the real axis of t and the circuits around the points B and D, we readily establish the required correspondence.

Turning now to the inclined sheetpile shown in Fig. 5-11a, we see that the flow domain contains only the slit BCD (Fig. 5-11b). Applying the Schwarz-Christoffel transformation, we have for the mapping of the

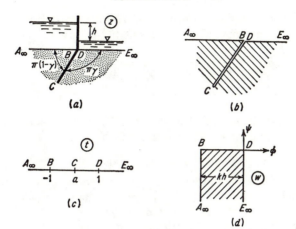

Fig. 5-11

region $A_\infty BCDE_\infty$ onto the lower half of the t plane (Fig. 5-11c), with the points B, C, and D going into the points $t = -1$, $t = a$, and $t = +1$,

$$z = M \int \frac{dt}{(t+1)^\gamma (t-a)^{-1} (t-1)^{1-\gamma}} + N \qquad (2)$$

Equation (1) for this case simply becomes

$$z = C(1+t)^{1-\gamma}(1-t)^\gamma \qquad (3)$$

Since Eqs. (2) and (3) must be equal, it follows that after differentiation

$$\frac{dz}{dt} = M_1(1+t)^{-\gamma}(t-a)(1-t)^{\gamma-1}$$
$$= -C(1+t)^{-\gamma}(1-t)^{\gamma-1}(t+2\gamma-1) \qquad (4)$$

Hence $M_1 = -C$, and $a = 1 - 2\gamma$.

Taking s as the length of the sheetpile and noting that its tip corresponds to $t = a$, we have

$$z = se^{-\pi\gamma i}$$

which, after substitution into Eq. (3), yields

$$C = \frac{se^{-\pi\gamma i}}{(1+a)^{1-\gamma}(1-a)^\gamma}$$

Now, substituting the expression for C into Eq. (3), we obtain for the mapping of the z plane onto the t plane,

$$z = se^{-\pi\gamma i}\left(\frac{1+t}{1+a}\right)^{1-\gamma}\left(\frac{1-t}{1-a}\right)^\gamma \qquad (5)$$

where $a = 1 - 2\gamma$.

Finally, noting that the w plane is a semi-infinite strip (Fig. 5-11d), we have for the mapping of the w plane onto the t plane,

$$\frac{dw}{dt} = \frac{kh}{\pi \sqrt{t^2 - 1}} \tag{6}$$

and

$$t = \cos \frac{\pi w}{kh} \tag{7}$$

The nature of the exit gradient (at point D) as a function of γ can be obtained directly from Eq. (2), Sec. 5-1 (cf. Prob. 7).

5-4. Finite Lower Impervious Boundary: General*

In Sec. 5-1 it was shown that the complex potential for a confined flow region bounded from below by an impervious boundary would lie within a rectangle in the w plane. Thus the mapping of the interior of a rectangle onto a half plane is of particular importance.

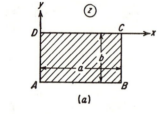

(a)

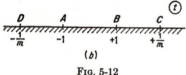

(b)

Fig. 5-12

Let us consider the mapping of rectangle $ABCD$ of Fig. 5-12a onto the lower half of the t plane (Fig. 5-12b). The correspondence between points is as shown on the figure, and m is some number less than unity, the value of which is to be determined. Applying the Schwarz-Christoffel transformation with all interior angles equal to $\pi/2$, we obtain

$$z = M \int_0^t \frac{dt}{\sqrt{(1 - t^2)(1 - m^2 t^2)}} + N \tag{1}†$$

Now, noting that $z = -ib + a/2$ corresponds to $t = 0$ (from symmetry), we find immediately that $N = -ib + a/2$. For points B, where $z = a - ib$ and $t = 1$, Eq. (1) becomes

$$\frac{a}{2} = M \int_0^1 \frac{dt}{\sqrt{(1 - t^2)(1 - m^2 t^2)}}$$

which, since the integral is the complete elliptic integral of the first kind (K), yields $M = a/2K$. Hence Eq. (1) can be written as

$$z = \frac{a}{2K} \int_0^t \frac{dt}{\sqrt{(1 - t^2)(1 - m^2 t^2)}} + \frac{a}{2} - ib \tag{2}$$

---

* A review of Appendix B through Sec. 3 is recommended before proceeding with this section.

† The integration implied by the Schwarz-Christoffel transformation is, in effect, a line integration along the real axis of the t plane.

At points C, $z = a$ and $t = 1/m$; Eq. (2) becomes

$$\frac{a}{2} + ib = \frac{a}{2K} \int_0^{1/m} \frac{dt}{\sqrt{(1 - t^2)(1 - m^2 t^2)}}$$

which, since the integral is $K + iK'$, yields finally

$$\frac{b}{a} = \frac{K'}{2K} \tag{3}$$

Thus, once the sides a and b of the rectangle are known, the modulus m can be obtained from tables of K'/K (cf. Table B-1, Appendix B), or conversely, knowing the modulus, one can find the ratio b/a.

Considering the section of dam (weir or spillway) shown in Fig. 5-13a, we shall specify as before that $\phi = 0$ along the tail-water boundary and

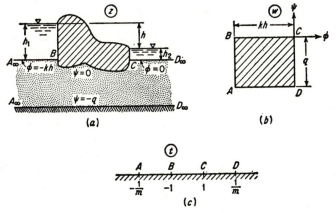

Fig. 5-13

$\phi = -kh$ along the headwater boundary· also, along the base of the structure $\psi = 0$, and along the impervious boundary $\psi = -q$, where q is the discharge per unit length of the structure. The resulting w plane is shown in Fig. 5-13b. Following the procedure outlined above, we have, for the mapping of the w plane onto the t plane (Fig. 5-13c),

$$w = -\frac{kh}{2K} \left[K - \int_0^t \frac{dt}{\sqrt{(1 - t^2)(1 - m^2 t^2)}} \right] \tag{4}$$

Now, noting that $u = \int_0^t [(1 - t^2)(1 - m^2 t^2)]^{-\frac{1}{2}} dt$ can be written as $t = \mathrm{sn}\, u$, Eq. (4) becomes

$$t = \mathrm{sn}\left(K + \frac{2Kw}{kh} \right) \tag{5}$$

which provides the required transformation between the w plane and the t plane subject to the determination of the modulus m. Generally, m is determined from characteristic lengths in the z plane [cf. Eq. (2), Sec. 5-2].

Considering the correspondence between points D, where $w = -iq$ and $t = 1/m$, we find, from either Eq. (4) or Eq. (5), that the discharge through the section is

$$q = \frac{khK'}{2K} \tag{6}$$

Comparing Eq. (6) with Eq. (4), Sec. 1-12, we see that $K'/2K$ for this case is equivalent to the ratio of the number of flow channels N_f to the number of equipotential drops N_e obtained from the graphical flow net solution.

5-5. Impervious Structure with Sheetpile on Layer of Finite Depth

As in Sec. 5-2 we shall again consider a general class of problems. The section to be investigated is shown in Fig. 5-14a.* Here again, once

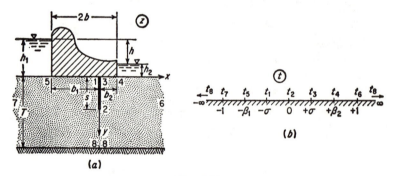

Fig. 5-14

the general problem has been solved, particular solutions can be obtained by adjusting the various dimensions of the section. This problem was solved, independently, by Pavlovsky [110] in 1922 and Muskat [99] in 1936. In the present development we shall follow more closely Pavlovsky's method of solution.

Choosing the correspondence between points in the z and t planes as shown on the figure, we have from the Schwartz-Christoffel transformation

$$z = M \int_0^t \frac{t\,dt}{(1 - t^2)\sqrt{\sigma^2 - t^2}} + N \tag{1}$$

* In this problem it is more convenient to take the width of the structure as $2b$.

Making the substitutions

$$\sigma^2 - t^2 = \tau^2 \qquad t \, dt = -\tau \, d\tau$$

and

$$1 - t^2 = \tau^2 + \sigma'^2$$

where $\sigma'^2 = 1 - \sigma^2$, we obtain

$$z = -\frac{M}{\sigma'}\left(\tan^{-1}\frac{\sqrt{\sigma^2 - t^2}}{\sigma'} - \tan^{-1}\frac{\sigma}{\sigma'}\right) + N \tag{2}$$

At points 2, $z = is$ and $t = 0$; from Eq. (1) we find $N = is$.
At points 3, $z = 0$ and $t = \sigma$; we have from Eq. (2)

$$\frac{M}{\sigma'}\tan^{-1}\frac{\sigma}{\sigma'} + is = 0 \tag{3}$$

and hence Eq. (2) reduces to

$$z = -\frac{M}{\sigma'}\tan^{-1}\frac{\sqrt{\sigma^2 - t^2}}{\sigma'} \tag{4}$$

where M and σ are still to be determined. From points 8, where $z = iT$ and $t = \infty$, we find, substituting into Eq. (4), that $M = -2iT\sigma'/\pi$ and hence

$$t = \pm\sqrt{\sigma^2 + \sigma'^2 \tanh^2\frac{\pi z}{2T}} \tag{5}*$$

Now, substituting for M into Eq. (3), we have

$$\frac{2Ti}{\pi}\tan^{-1}\frac{\sigma}{\sigma'} = is$$

whence

$$\sigma = \sin\frac{\pi s}{2T} \qquad \sigma' = \cos\frac{\pi s}{2T} \tag{6}$$

and the required transformation between the z and t planes is

$$t = \pm\cos\frac{\pi s}{2T}\sqrt{\tan^2\frac{\pi s}{2T} + \tanh^2\frac{\pi z}{2T}} \tag{7}\dagger$$

In particular, for the ends of the structure, substituting $z = -b_1$ and $z = b_2$ into Eq. (7), we obtain

$$\beta_2 = \cos\frac{\pi s}{2T}\sqrt{\tanh^2\frac{\pi b_2}{2T} + \tan^2\frac{\pi s}{2T}}$$

$$\beta_1 = \cos\frac{\pi s}{2T}\sqrt{\tanh^2\frac{\pi b_1}{2T} + \tan^2\frac{\pi s}{2T}} \tag{8}$$

---

* The plus sign applies to the right half of the t plane, the minus sign to the left half.
† *Ibid.*

A plot of Eqs. (8) is given in Fig. 5-15. The figure can also be used for real values of z/T in Eq. (7), that is, taking $z/T = b/T$ and $t = \beta$. For example, if $s/T = 0.25$ and $z/T = 0.4$, $\beta = t = 0.64$.

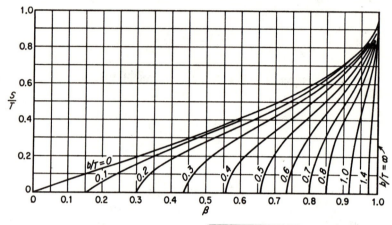

FIG. 5-15. $\beta = \cos \dfrac{\pi s}{2T}\sqrt{\tanh^2 \dfrac{\pi b}{2T} + \tan^2 \dfrac{\pi s}{2T}}$

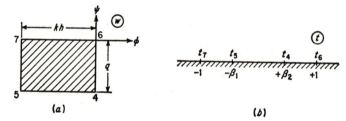

FIG. 5-16

Turning now to the mapping of the w plane onto the t plane (Fig. 5-16), we have, from the Schwarz-Christoffel transformation,

$$w = M_1 \int_0^t \frac{dt}{\sqrt{(t - 1)(t - \beta_2)(t + \beta_1)(t + 1)}} + N_1 \qquad (9)$$

Now, separating Eq. (9) into the sum of the two integrals,

$$w = M_1 \int_0^1 (\quad) + M_1 \int_1^t (\quad) + N_1$$

and noting at points 6 that $w = 0$ and $t = 1$, Eq. (9) reduces to

$$w = M_1 \int_1^t \frac{dt}{\sqrt{(t - 1)(t - \beta_2)(t + \beta_1)(t + 1)}} \qquad (10)$$

Performing the indicated integration (cf. Eq. 550, Ref. [111]), we obtain

$$\sqrt{\frac{(t-1)(1+\beta_2)}{2(t-\beta_2)}} = \operatorname{sn} \frac{\mu w}{M_1} \tag{11a}$$

or
$$t = \frac{\beta_2 + 1 - 2\beta_2 \operatorname{sn}^2 (\mu w/M_1)}{\beta_2 + 1 - 2 \operatorname{sn}^2 (\mu w/M_1)} \tag{11b}$$

where
$$\mu = \frac{1}{2} \sqrt{(1+\beta_1)(1+\beta_2)}$$

$$\text{modulus } m = \sqrt{\frac{2(\beta_1 + \beta_2)}{(1+\beta_1)(1+\beta_2)}}$$

To evaluate the constant M_1: at points 5, $t = -\beta_1$ and $w = -iq - kh$; hence, substituting these values into Eq. (11a), we have

$$\operatorname{sn}\left(\frac{\mu w}{M_1}\right) = \frac{1}{m}$$

or
$$w = -iq - kh = \frac{M_1}{\mu}(K + iK')$$

whence
$$M_1 = -\frac{\mu kh}{K} \qquad \text{and} \qquad q = \frac{khK'}{K} \tag{12}$$

Thus the required transformation between the w and t planes is

$$t = \frac{\beta_2 + 1 - 2\beta_2 \operatorname{sn}^2 (Kw/kh)}{\beta_2 + 1 - 2 \operatorname{sn}^2 (Kw/kh)} \tag{13}$$

with modulus m as given in Eq. (11b).

For a symmetrically placed sheetpile, $\beta = \beta_1 = \beta_2$; hence, making use of the relationship

$$\frac{K'(m)}{K(m)} = \frac{K'(\beta)}{2K(\beta)} \tag{14a}$$

where β is given by Eq. (7)

$$\beta = \cos \frac{\pi s}{2T} \sqrt{\tanh^2 \frac{\pi b}{2T} + \tan^2 \frac{\pi s}{2T}} \tag{14b}*$$

we obtain for the discharge

$$q = \frac{khK'(\beta)}{2K(\beta)} \tag{14c}$$

A plot giving the discharge (q/kh) for symmetrically placed pilings as a function of the depth of embedment (s/T) and the width of the structure (b/T) is shown in Fig. 5-17. Figure 5-18 illustrates the influence of the

* See Fig. 5-15.

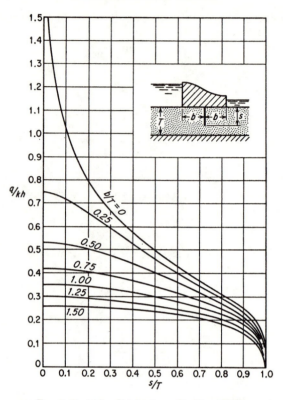

Fig. 5-17. (*After Polubarinova-Kochina* [116].)

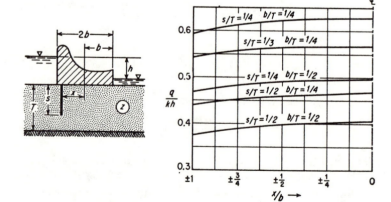

Fig. 5-18

position of the piling on the discharge for various combinations of depth of embedment and size of structure. It is apparent from Fig. 5-18 that although the discharge is a maximum when the sheetpile is at the center of the structure, the variation with position is small (Fig. 5-19). Hence

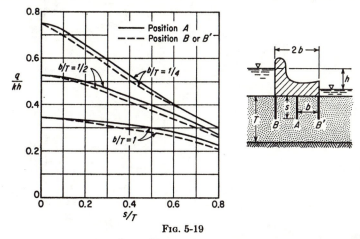

Fig. 5-19

for design purposes the magnitude of the discharge for any position of the piling may be taken as that with the piling set at the center (Fig. 5-17).

Figure 5-17 demonstrates that the quantity of seepage decreases as either the depth of the sheetpile or the width of the weir increases; however the benefit to be gained by increasing the depth of sheetpile embedment is seen to decrease sharply as the ratio of the width of the weir to the thickness of the permeable layer increases. Indeed, for half-widths larger than the thickness, the quantity of seepage is seen to be almost independent of the depth of embedment unless the sheetpile extends close to the base of the permeable layer. Thus little or no material advantage is to be gained by increasing the piling depth for ratios of $b/T \geqq 1$ unless the piling can be driven into the impervious base. This is particularly noteworthy, as with increasing depths of driving the risk of faulty connections between the individual piling sections is also likely to increase.

Finally, to obtain the discharge for other than flat-bottom structures, we note that in Sec. 5-2 any streamline can be taken as an impervious boundary. Hence, recalling that $q = \Delta\psi$, by defining $\psi = 0$ along the base of the structure, one may obtain a good approximation to the discharge quantity by assigning $\psi = -q$ to any streamline in the vicinity of the impervious boundary. The rather insensitive nature of K'/K obviates the need for greater refinement.

The distribution of uplift pressures p acting on the base of a structure without piling is shown in Fig. 5-20. We note from this figure that for

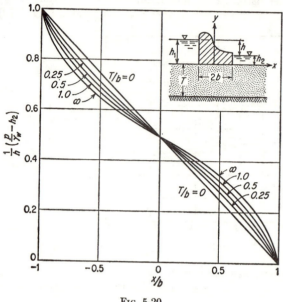

Fig. 5-20

T/b ratios greater than unity the difference between the exact solution and the approximation of an infinite depth of porous media ($T/b = \infty$) is negligible and on the *safe side* for the higher pressures.

The question of the pressures acting on the bases of weirs with and without piling was investigated by Pavlovsky in some detail. The curves of Fig. 5-21 were prepared to show the pressure drop (Δp) across pilings as a percentage of the total pressure drop across the weir for various depths of embedment. Two cases are considered: (1) a sheetpile at the center, and (2) a sheetpile at the heel of the structure. On the basis of these curves, Pavlovsky concluded that in the case of central sheetpiles, for ratios of $T/b < 2$, the benefit to be gained from the pile increases with increased depth of embedment, whereas for ratios of $T/b > 2$ this advantage decreases. Thus it appears that a depth of embedment beyond $s/b \approx 1$ is not warranted in the case of centrally placed pilings unless complete embedment can be achieved at relatively shallow depths. This criterion is particularly advised since the risk of faulty joints also increases with increased depth of driving. For sheetpiles at the heel the critical thickness is shown to be $T/b < \frac{1}{2}$, and the recommended maximum depth of driving is $s/b \approx 0.5$.

Finally, it should be noted in Fig. 5-21 that the assumption of an infinite depth of permeable soil would result in an approximation for the pressure

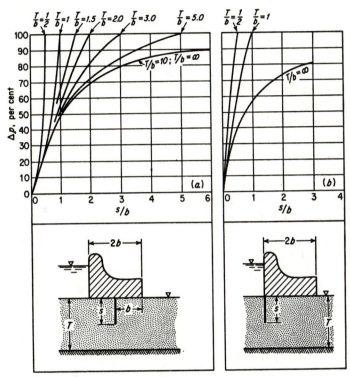

FIG. 5-21. (*After Pavlovsky* [110].)

drop that would be on the safe side. Whereas the deviations are shown to be greater for the shallower depths, for these the piling would probably be driven full depth. Thus there exists a well-grounded justification for using the simpler infinite-depth solutions previously developed in Sec. 5-2 for the uplift pressures acting on the bases of structures with or without piling.

To obtain the exit gradient with the sheetpile at the toe of the structure (at point 3 = point 4 in Fig. 5-14a), again using Eq. (2), Sec. 5-1, we find that

$$I_E = \frac{h\pi}{4KT\sigma} \sqrt{\frac{2\sigma(1 + \beta_1)(1 + \sigma)}{\sigma + \beta_1}} \tag{15a}$$

where the modulus is

$$m = \sqrt{\frac{2(\beta_1 + \sigma)}{(1 + \beta_1)(1 + \sigma)}} \tag{15b}$$

and $\sigma = \sin(\pi s/2T)$. Now, letting point 5 approach point 1 (Fig. 5-14), we have the solution for a single sheetpile for which Eqs. (15) reduce to

$$I_E = \frac{h\pi}{4KTm} \tag{16a}$$

where the modulus is now

$$m = \sigma = \sin\frac{\pi s}{2T} \tag{16b}$$

Substituting $K_{\min} = \pi/2$ and $\sin(\pi s/2T) = \pi s/2T$, we find for the maximum exit gradient the same value obtained previously (Fig. 5-9c) from the approximate theory,

$$I_{E\max} = \frac{h}{\pi s} \tag{17}$$

Hence it is apparent that for exit gradients also the assumption of an infinite depth of porous media will be on the safe side. A similar state of affairs can be shown for a structure with a single sheetpile by comparing Eqs. (15) with the expression given in Fig. 5-9d. In Fig. 5-22 is shown a

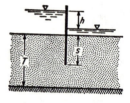
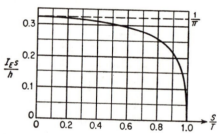

FIG. 5-22

plot of $I_E s/h$ as a function of s/T as given in Eqs. (16). The approximate theory yields for this case $1/\pi$. It is apparent from this figure that not only is the approximate value on the safe side but that the error is of negligible order for ratios of $s/T < 0.75$ (which will generally be within the order of embedment unless the piling is anchored into the base).

5-6. Heaving and Roofing

Although it is commonly accepted that the exit gradient will generally provide the most significant design criterion for the factor of safety with respect to piping (cf. Art. 59, Ref. 145 and page 78, Ref. 86), some consideration should also be given to the possibility of piping originating within the soil mass (called *heaving*).

On the basis of his model tests, Terzaghi [144] found that, for a single row of sheetpiles of length s, the stability with respect to heaving could be evaluated by considering the average hydraulic gradient across the prism of soil of depth s and width $s/2$, as shown (crosshatched) in Fig.

5-23a. That is, the factor of safety with respect to heave is

$$\text{F.S.} = \frac{sk I_{cr}}{\phi_{av}} \tag{1}$$

where ϕ_{av} = average potential difference across width $s/2$ (Fig. 5-23a)
k = coefficient of permeability
I_{cr} = critical gradient [Eqs. (8) and (9), Sec. 1-13]

The distribution of ϕ, and hence the value of ϕ_{av}, can be obtained directly from the work of Sec. 5-2. For sections other than the single row of sheetpiles (such as Fig. 5-23b), Terzaghi recommended that the stability

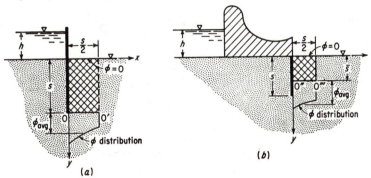

Fig. 5-23

be investigated at various horizontal sections ($y = \bar{s} < s$) to determine the minimum factor of safety. Generally a factor of safety as per Eq. (1) (with $s = \bar{s}$) of 4 to 5 is considered adequate.* If the factor of safety is too small, it may be increased by placing an inverted filter† of height H on the surface of the soil. For this, the factor of safety will be given by

$$\text{F.S.} = \frac{\bar{s}k I_{cr}}{\phi_{av}} + \frac{Hk\gamma_f}{\phi_{av}\gamma_w} \tag{2}$$

where γ_f is the effective unit weight of the filter material.

An empirical approach to the problem of piping was developed by Lane [82] based upon his observations of over 200 masonry dams founded on soils. Lane recognized that in order to prevent piping, the velocity of the water must be insufficient to cause the removal of the foundation material on the downstream side of the dam. He reasoned that, since the resistance to seepage may be much less along the base of the structure than along other streamlines (due to the difficulty of achieving positive contact between the bottom of the masonry and the top of the foundation material), the path of this upper flow line represented the critical path

---

* As an approximation that will generally be sufficient, ϕ_{av} can be taken as the ϕ value along the piling at $y = s$.
† See Art. 11, Ref. 145.

from the standpoint of failure due to piping.   The path of this upper flow line is called the *creep path* and the phenomenon associated with this type of piping is called *roofing*.

From his analysis of the action of actual dams, Lane concluded that the creep path along contact surfaces with slopes less than 45° from the horizontal offer only one-third the resistance to roofing as do those with slopes greater than 45°.   On the basis of this criterion he developed the equation for the *weighted creep ratio* R_c:

$$R_c = \frac{\frac{1}{3}H + V}{h} \qquad (3)$$

where H = horizontal contacts ($\leq 45°$)
V = vertical contacts ($>45°$)
h = head loss through the system

For the section illustrated in Fig. 5-24, the creep path is $BCDEFG$, $H = 66$ ft, $V = 83$ ft, and $h = 15$ ft; hence the weighted creep ratio is

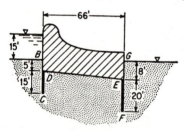

Fig. 5-24

$R_c = 7.0$.   For safety Lane recommended that R_c should not be less than the values given in Table 5-2 for the particular foundation material.

Table 5-2. Recommended Weighted Creep Ratios*

Material	*Safe weighted creep-head ratios, Rc*
Very fine sand or silt	8.5
Fine sand	7.0
Medium sand	6.0
Coarse sand	5.0
Fine gravel	4.0
Medium gravel	3.5
Coarse gravel, including cobbles	3.0
Boulders with some cobbles and gravel	2.5
Soft clay	3.0
Medium clay	2.0
Hard clay	1.8
Very hard clay or hardpan	1.6

* From E. W. Lane, Security from Under-seepage: Masonry Dams on Earth Foundations, *Trans. Am. Soc. Civil Eng.*, vol. 100, p. 1257, 1935.

We see from this table that the section of Fig. 5-24, according to Lane, would be safe for all materials except very fine sand or silt.*

The values given in Table 5-2 are recognized as being very conservative, tending toward maximum rather than average values. Not a single failure was found by Lane where the dam evidenced a weighted creep-head ratio as large as those given.

Whereas roofing is a consideration only for those cases where the possibility of imperfect contact exists between the base of the structure and the surface of the foundation material, provisions which would ensure this contact, such as grouting under the structure, are strongly recommended.

Blind application of any piping criterion without cognizance of subsurface soil conditions can lead to dangerous results. For example, for the condition illustrated in Fig. 5-25a, the critical location for piping will

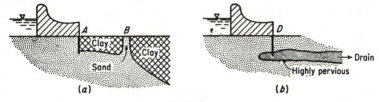

Fig. 5-25

probably not occur at point A but at point B, whereas for the subsurface condition in Fig. 5-25b, it is highly probable that the tail water may never reach point D. Finally, with respect to Lane's criterion, we note that although the weighted creep ratios of both sections in Fig. 5-26 may be

Fig. 5-26

identical, the design shown in (b) would have a high probability of failure due to incipient piping in the vicinity of point C.

In summary, the following procedure is recommended (assuming adequate exploration of the subsurface) for design considerations with respect to piping:

---

* R_c, in Eq. (3), is somewhat equivalent to the reciprocal of the gradient. Hence, assuming the critical gradient is approximately equal to unity, the values in Table 5-2 can be thought of as quasi-factors-of-safety with respect to piping.

1. Design the section on the basis of the exit gradient criterion, providing sound cutoffs of either masonry or piling (for shallow depths). Where feasible consider the economy of anchoring cutoffs in an impervious base regardless of the design criterion.

2. Check the factor of safety with respect to the heaving of the deeper layers (Terzaghi's criterion).

3. Consider the use of reverse filters on the downstream surface or where required.*

4. Take advantage of impervious fill (or blankets) on the upstream surface to increase seepage distance.

5-7. Depressed Structure on a Permeable Base of Infinite Extent.†

Throughout the previous sections of this chapter it was assumed that the base of the structure rested on the surface of the flow medium. We shall now investigate the effects of setting the structure below the surface.‡ The first section to be considered and the correspondence between points are shown in Fig. 5-27.

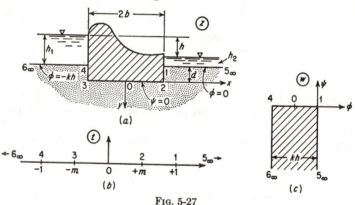

FIG. 5-27

For the mapping of the z plane onto the t plane, we have

$$z = M \int_0^t \sqrt{\frac{m^2 - t^2}{1 - t^2}}\, dt + N \tag{1}$$

which, since $t = 0$ when $z = 0$, yields immediately $N = 0$.   Now, divid-

---

* Referring to reverse filters, Terzaghi and Peck [145, Art. 59] state: "The presence of the filters prevents even incipient erosion at all points of the protected area and increases the critical head from the value required to produce erosion to the much larger value required to produce failure by heave."

† Before proceeding with this section, a review of the elliptic integral of the second kind (cf. Sec. B-4) is recommended.

‡ The developments of the solution up to Eq. (4) and of the next section, in a somewhat modified form, were first obtained by Pavlovsky in 1922.

ing the numerator of the integrand by m^2, we obtain [Eq. (16a), Appendix B],

$$z = MmE\left(\frac{1}{m}, \phi\right) = M[E(m,\theta) - m'^2F(m,\theta)] \tag{2}$$

where $\theta = \sin^{-1}(\sin \phi/m) = \sin^{-1}(t/m)$ and $F(m,\theta)$ and $E(m,\theta)$ are the elliptic integrals of the first and second kind, respectively.

To determine the constants M and m, we consider the correspondence at points 1 and 2. At points 2, $z = b$ and $t = m$; thus $\theta = \pi/2$, $F(m,\theta) = K$, $E(m,\theta) = \mathbf{E}$ and

$$M = \frac{b}{\mathbf{E} - m'^2K} \tag{3}$$

At points 1, $z = b - id$ and $t = 1$; thus $\theta = \sin^{-1}(1/m)$,

$$F\left(m, \frac{1}{m}\right) = K + iK'$$

and $E(m,1/m) = \mathbf{E} + i(K' - \mathbf{E}')$. Substituting into Eq. (2), we find

$$\frac{d}{b} = \frac{\mathbf{E}' - m^2K'}{\mathbf{E} - m'^2K} \tag{4}$$

A plot of Eq. (4) which yields m as a function of d/b (or b/d) is given in Fig. 5-28.

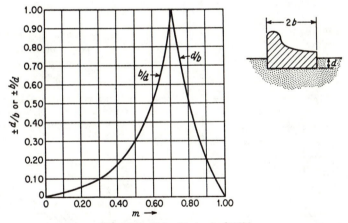

Fig. 5-28. (*After Pavlovsky* [110].)

Substituting for M into Eq. (2), we have, finally, for the mapping of the z plane onto the t plane,

$$z = \frac{b[E(m,\theta) - m'^2F(m,\theta)]}{\mathbf{E} - m'^2K} \tag{5}$$

where m can be obtained from Fig. 5-28 and $\theta = \sin^{-1}(t/m)$.

For the mapping of the w plane onto the t plane we have from Eq. (5), Sec. 4-7,

$$t = \cos \frac{\pi w}{kh} \qquad (6)$$

Hence the pressure in the water is

$$p = \gamma_w \left(\frac{h}{\pi} \cos^{-1} t + y + h_2 + d \right) \qquad 0 \leqq \cos^{-1} t \leqq \pi \qquad (7)$$

Because of the implicit nature of the above transformations, the determination of the pressure distribution acting on the base structure can best be achieved by first selecting a value of t ($-m \leqq t \leqq m$ under the floor, and $m < |t| < 1$ along the vertical sides) and then finding the corresponding pressure from Eq. (7) and the corresponding point in the z plane from Eq. (5). In particular, at points 2 and 3 of Fig. 5-27, where $t = +m$ and $t = -m$, respectively, we have

$$p = \gamma_w \left[\frac{h}{\pi} \cos^{-1}(\pm m) + h_2 + d \right] \qquad (8)$$

Applying Eq. (2), Sec. 5-1, we find without difficulty, for the exit gradient (point 1),

$$I_E = \frac{h}{\pi d} \frac{E' - m^2 K'}{m'} \qquad (9)$$

A plot of $(E' - m^2 K')/m'$ as a function of the modulus m is given in Fig. 5-29. Combining this figure with Fig. 5-28, we easily obtain the exit gradient. The method of solution will now be illustrated by an example.

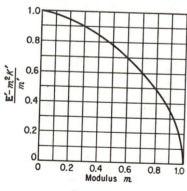

FIG. 5-29

Example 5-2. For the section of Fig. 5-30 determine (a) the factor of safety with respect to piping, and (b) the pressure distribution along the floor of the weir.

Here, $b = 20$ ft and $d = 5$ ft; hence from Fig. 5-28, with $d/b = 0.25$, we find the

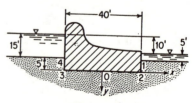

FIG. 5-30

modulus $m = 0.88$. With $m = 0.88$, from Fig. 5-29 we find $(E' - m^2 K')/m' = 0.36$, and hence from Eq. (9) we have the exit gradient $I_E = 0.23$. Considering a critical gradient of unity, we would have a factor of safety with respect to piping (exit-gradient criterion) of approximately 4. Referring to Khosla's recommendations for factors of safety (Sec. 5-1), we see that the design would be barely safe in a foundation composed of gravel. Using Lane's method (Sec. 5-6), we have a weighted creep ratio of 2.33, and from Table 5-2 we find the design just shy of the recommended safe value of 2.5 for boulders with some cobbles and gravel.

For the pressure along the floor we have, from Eq. (8) at point 2 (Fig. 5-30),

$$p_2 = \gamma_w \left[\frac{10}{\pi} \cos^{-1} (0.88) + 10 \right] = 11.6\gamma_w$$

At point 0, $t = 0$; hence

$$p_0 = \gamma_w \left[\frac{10}{\pi} \cos^{-1} (0) + 10 \right] = 15.0\gamma_w$$

At point 3, $t = -m$; Eq. (8) yields

$$p_3 = \gamma_w \left[\frac{10}{\pi} \cos^{-1} (-0.88) + 10 \right] = 18.4\gamma_w$$

Note that p_3 could have been obtained directly from p_2; that is, $\cos^{-1} (-0.88) = \pi - \cos^{-1} (0.88)$.

Let us obtain the pressure corresponding to $t = \pm m/2$. At $t = +m/2$, from Eq. (7) we have

$$p_{t=m/2} = \gamma_w \left[\frac{10}{\pi} \cos^{-1} (0.44) + 10 \right] = 13.54\gamma_w$$

and hence for $t = -m/2$,

$$p_{t=-m/2} = 16.46 \gamma_w$$

The z coordinates on the floor, corresponding to $t = \pm m/2$, are determined from Eq. (5). For this case, with $\theta = \sin^{-1} (t/m) = 30°$ and $m = 0.88$, we obtain $x = \pm 10.9$ ft. The resulting pressure distribution is shown in Fig. 5-31 (solid line). The superimposed dotted line represents the pressure distribution, assuming the weir rests on the surface ($d = 0$) with an additional pressure, due to elevation head, of $10\gamma_w$ (Fig. 4-11).

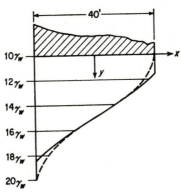

FIG. 5-31

We note in Fig. 5-31 that the actual distribution of pressure (solid line) varies but little from a straight line, being slightly above on the upstream side and below by the same amount on the downstream portion. A somewhat similar relationship is observed between the exact solution and the approximation (dotted line) for the structure assumed to be resting on the surface (considering pressures due to elevation head separately). Studies conducted by Pavlovsky with various ratios of $d/2b$ ($d/2b = 0$, $\frac{1}{10}$, $\frac{1}{5}$, $\frac{1}{3}$, 1.0, 3.0, 5.0) indicate that the difference between the pressures computed from the exact theory and the above approximation for $x \leqq \pm 3b/4$ will be less than 12 per cent for $d/2b = \frac{1}{5}$. For $d/2b = \frac{1}{3}$, the difference will be less than 15 per cent.

5-8. Depressed Structure on Permeable Base of Infinite Extent with Two Symmetrical Rows of Pilings

The correspondence between the z plane and the t plane is shown in Fig. 5-32. Hence

$$z = M \int_0^t \frac{(t^2 - \sigma^2)\, dt}{\sqrt{(1 - t^2)(m^2 - t^2)}} \tag{1}$$

Writing the numerator of the integrand as $(t^2 - m^2) + (m^2 - \sigma^2)$, we can split the integral into

$$z = -M \int_0^t \sqrt{\frac{m^2 - t^2}{1 - t^2}}\, dt + \frac{M(m^2 - \sigma^2)}{m} \int_0^t \frac{dt}{\sqrt{(1 - t^2)(1 - t^2/m^2)}}$$

Recognizing the first integral to be the same as Eq. (1), Sec. 5-7, and the

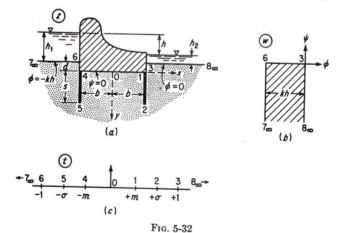

Fig. 5-32

second integral as the elliptic integral of the first kind of modulus $1/m$, we obtain

$$z = -M[(\sigma^2 - 1)F(m,\theta) + E(m,\theta)] \tag{2}$$

where $\theta = \sin^{-1}(t/m)$. We note that, using Jacobi's notation, Eq. (2) can also be written as

$$z = -M[(\sigma^2 - 1)u + E(u)] \tag{3}$$

where sn $u = t/m$.

To determine the unknown constants M, σ, and m, we consider the relationship between the points 1, 2, and 3 in the z and t planes.

At points 1, $z = b$ and $t = m$; hence sn $u = t/m = 1$, $u = K$, $E(u) = E(K) = \mathbf{E}$, and

$$b = -M[(\sigma^2 - 1)K + \mathbf{E}] \tag{4}$$

At points 2, $z = b + is$ and $t = \sigma$; hence sn $u = \sigma/m > 1$ but $< 1/m$, which demonstrates that u for this case will be a complex number of the form $u = K + iv$ [cf. Eqs. (12), Sec. B-3]. **Thus**

$$\frac{\sigma}{m} = \text{sn}\ (K + iv) = \frac{1}{\text{dn}\ (v, m')}$$

and $v = \text{dn}^{-1}(m/\sigma, m')$ [Eq. (12a), Appendix B]. From Eq. (19b), Appendix B we have

$$E(u) = E(K + iv) = \mathbf{E} + i[v - E(m', v) + m'^2\ \text{sn}\ v'\ \text{cn}\ v'/\text{dn}\ v']^*$$

hence for points 2 we obtain

$$b + is = -M[(\sigma^2 - 1)K + \mathbf{E}] \\ - Mi\left[\sigma^2 v + \frac{m'^2\ \text{sn}\ v'\ \text{cn}\ v'}{\text{dn}\ v'} - E(m', v)\right] \tag{5}$$

At points 3, $z = b - id$ and $t = 1$; hence sn $u = 1/m$, $u = K + iK'$, $E(u) = E(K + iK') = \mathbf{E} + i(K' - \mathbf{E}')$, and

$$b - id = -M[(\sigma^2 - 1)K + \mathbf{E}] - Mi(\sigma^2 K' - \mathbf{E}') \tag{6}$$

On the basis of the above, we find that the three constants M, m, and σ can be determined from the equations

$$M[(\sigma^2 - 1)K + \mathbf{E}] = -b \tag{7a}$$

$$M\left[\sigma^2 v + \frac{m'^2\ \text{sn}\ v'\ \text{cn}\ v'}{\text{dn}\ v'} - E(m', v)\right] = -s \tag{7b}$$

$$M(\sigma^2 K' - \mathbf{E}') = d \tag{7c}$$

---

* sn v', cn v', and dn v' designate sn (v, m'), cn (v, m'), and dn (v, m').

Combining Eqs. (7a) and (7c), we have for σ^2,

$$\sigma^2 = \frac{\mathbf{E}' + (d/b)(K - \mathbf{E})}{K' + (d/b)K} \tag{8}$$

and hence σ can be determined once m is known. Substituting $(\sigma^2 - 1)$ from Eq. (8) into Eq. (7a), we obtain

$$-\frac{M}{b} = \frac{K' + (d/b)K}{\mathbf{E}'K + \mathbf{E}K' - KK'}$$

which, recognizing Legendre's formula, $\mathbf{E}'K + \mathbf{E}K' - KK' = \pi/2$, yields

$$M = -\frac{2b}{\pi}\left(K' + \frac{d}{b}K\right) \tag{9}$$

Substituting Eq. (7a) for M into Eq. (7b), we find

$$\frac{s}{b} = \frac{\sigma^2 v + (m'^2 \operatorname{sn} v' \operatorname{cn} v'/\operatorname{dn} v') - E(m',v)}{(\sigma^2 - 1)K + \mathbf{E}} \tag{10}$$

The R.H.S. of this expression was shown to be a function of the modulus and the ratio of d/b only. Hence this expression can be plotted to yield the modulus as a function of the ratios s/b and d/b. Such a plot was obtained by Harr and Deen [53] and is given in Fig. 5-33. We note in Eq. (8) that when $s = 0$, $m = \sigma$ and

$$\frac{d}{b} = \frac{\mathbf{E}' - m^2 K'}{\mathbf{E} - m'^2 K} \tag{11}$$

which is precisely the expression we obtained in Sec. 5-7 [Eq. (4)] for the depressed structure without pilings.

Finally, substituting for M and $(\sigma^2 - 1)$ into Eq. (3), we obtain for the required transformation between the z and t planes,

$$z = \frac{2b}{\pi}\left[\left(\mathbf{E}' - K' - \frac{d}{b}\mathbf{E}\right)u + \left(K' + \frac{d}{b}K\right)E(u)\right] \tag{12}$$

where the modulus m is given in Fig. 5-33.

For the mapping of the w plane (Fig. 5-32) onto the t plane, we have again [Eq. (6), Sec. 5-7],

$$t = \cos\frac{\pi w}{kh}$$

Hence the pressure in the water is

$$p = \gamma_w\left(\frac{h}{\pi}\cos^{-1} t + y + h_2 + d\right) \qquad 0 \le \cos^{-1} t \le \pi \tag{13}$$

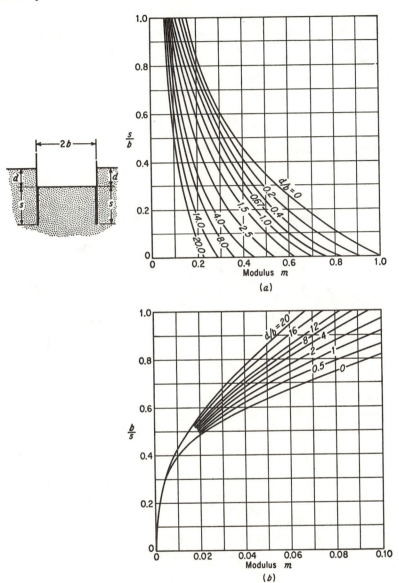

FIG. 5-33

Once again it is advisable to use an indirect approach (assuming t and finding the corresponding z) for the determination of the pressure distribution along the contour of the structure (see Example 5-2).

For the exit gradient (point 3), using Eq. (2), Sec. 5-1, we readily obtain

$$I_E = \frac{hm'}{2b\left(K' - E' + \dfrac{d}{b}\mathbf{E}\right)} \tag{14}$$

where the modulus is as is given in Fig. 5-33.

5-9. Double-wall Sheetpile Cofferdam

Figure 5-34 represents a section through a double-wall cofferdam consisting of two rows of sheetpiles. After the sheetpiles are driven, the

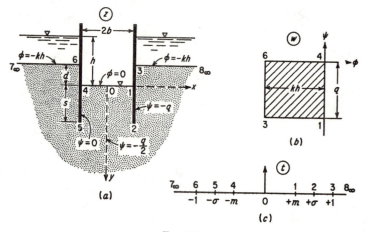

Fig. 5-34

soil between them is excavated to a depth d below the ground surface. We seek in this problem to determine the discharge quantity and the factor of safety with respect to piping.

Noting in Fig. 5-34 that the z plane and t plane are precisely the same as in Sec. 5-8, we have immediately for the required transformation between them [Eq. (12), Sec. 5-8],

$$z = \frac{2b}{\pi}\left[\left(\mathbf{E'} - K' + \frac{d}{b}\mathbf{E}\right)u + \left(K' + \frac{d}{b}K\right)E(u)\right] \tag{1}$$

where sn $u = t/m$, and the modulus m can be obtained directly from Fig. 5-33.

It is convenient in this problem to take the w plane as shown in Fig. 5-34b. Hence, for the mapping of the w plane onto the t plane, we have

$$w = \frac{M}{m} \int_0^t \frac{dt}{\sqrt{(1 - t^2)(1 - t^2/m^2)}} - \frac{iq}{2} = Mu - \frac{iq}{2} \tag{2}$$

where, as above, sn $u = t/m$.

Considering the correspondence at points 1, $t = m$ and $w = -iq$; hence sn $u = 1$, $u = K$ and

$$M = -\frac{iq}{2K} \tag{3}$$

At points 3, $t = 1$ and $w = -kh - iq$; hence sn $u = 1/m$, $u = K + iK'$, and

$$q = \frac{2khK}{K'} \tag{4}$$

A plot of Eq. (4) as a function of the modulus (Fig. 5-33) is given in Fig. 5-35.

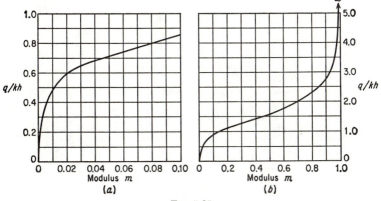

Fig. 5-35

Recalling that $q = kIA$, where A is the area of the section normal to the direction of flow, we find, for the average exit gradient along the bottom of the excavation,

$$I_{av} = \frac{q}{2kb} \tag{5}$$

For the determination of the maximum exit gradient along the base of the excavation (at points 1 and 4, $t = \pm m$), from Eq. (2), Sec. 5-1, we obtain the relation

$$I_E = I_1 = I_4 = \frac{h\pi}{2bK'[K' + (d/b)K](m^2 - c^2)} \tag{6}$$

where σ^2 is defined by Eq. (8), Sec. 5-8. A plot of Eq. (6) in terms of $I_E s/h$ is given in Fig. 5-36.

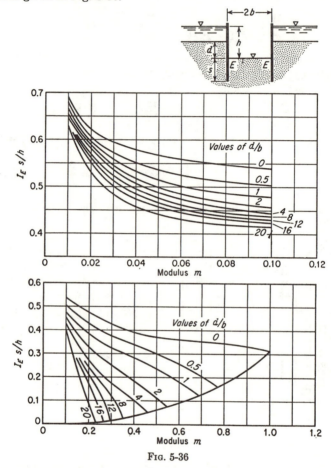

FIG. 5-36

Example 5-3. In Fig. 5-34, $h = 10$ ft, $d = 4$ ft, $2b = 40$ ft, and $s = 10$ ft. Determine (a) the reduced quantity of flow (q/k), (b) the average exit gradient, and (c) the maximum exit gradient.

From Fig. 5-33, with $s/b = 0.5$ and $d/b = 0.2$, we obtain the modulus $m = 0.35$. Then, from Fig. 5-35, we find $q/kh = 1.3$ and hence $q/k = 13$ ft.

From Eq. (5), we obtain the average gradient $I_{av} = \frac{13}{40} = 0.32$.

Next, entering Fig. 5-36 with $m = 0.35$ and $d/b = 0.2$, we find $I_E s/h = 0.39$, whence $I_E = 0.39$. Thus the factor of safety with respect to piping will be $1/0.39 \approx 2.6$.

PROBLEMS

1. Show that the transformation Eq. (3), Sec. 5-2, is valid for the points A and G of Fig. 5-2a.

2. Obtain the pressure distribution along all impervious boundaries in Fig. 5-4.

3. Obtain the general expression for the uplift force for a weir resting on the ground surface (of infinite depth) with a centrally placed sheetpile.

4. Demonstrate that each of the exit-gradient formulas in Fig. 5-9 can be obtained from Eq. (16), Sec. 5-2.

5. Verify that the complete mapping of the z plane onto the t plane in Fig. 5-10 is given by Eq. (1), Sec. 5-3.

6. Show that with $\gamma = 90°$, Eq. 5, Sec. 5-3, yields the transformation for a vertical sheetpile.

7. Derive the general expression for the exit gradient for an inclined sheetpile and discuss the nature of this gradient when γ in Fig. 5-11 is (a) equal to $\frac{1}{2}$, (b) less than $\frac{1}{2}$, and (c) greater than $\frac{1}{2}$.

8. Noting that the rectangle (w plane) in Fig. 5-13b becomes a semi-infinite strip as points A and D approach infinity, demonstrate that Eq. (5), Sec. 5-4, will degenerate into $t = \cos{(\pi w/kh)}$.

9. A 20 ft wide weir (without piling) rests on the surface of a 15-ft layer of soil. The head loss is 15 ft. Obtain the distribution of the factor of safety with respect to piping along the tail-water boundary for a distance of 50 ft downstream of the toe of the structure.

10. Verify Eqs. (14), Sec. 5-5.

11. Verify Eqs. (15), Sec. 5-5.

12. For the section shown in Fig. 5-37, obtain the general expression for (a) the quantity of seepage, (b) the exit gradient, and (c) the pressure distribution along the piling.

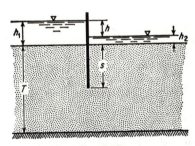

FIG. 5-37

13. Solve Prob. 3 for a layer of finite thickness T, and compare with the solution to Prob. 3.

14. For the section shown in Fig. 5-38, estimate the factor of safety with respect to (a) uplift force, (b) uplift moment (neglect moment due to piling), and (c) piping ($\gamma_m = 124.8$ pcf).

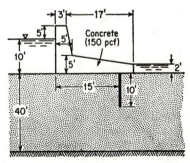

FIG. 5-38

15. For the section shown in Fig. 5-39, estimate (a) the length of impervious apron L required to reduce the pressure at point A to one-half of that without the apron, (b) the reduced discharge for part a, and (c) both the factor of safety with respect to piping and roofing, with and without the apron.

16. Neglecting the apron, obtain the factor of safety with respect to heave for the section shown in Fig. 5-39. Repeat for a depth of pile embedment of 15 ft and compare the ratios of the factors of safety of the heaving criterion to the exit-gradient criterion for the two depths of embedment.

17. Verify Eq. (9), Sec. 5-7.

18. In Fig. 5-30 find the depth of embedment required to double the factor of safety with respect to piping.

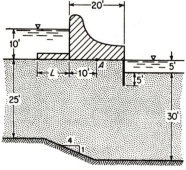

Fig. 5-39

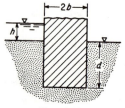

Fig. 5-40

19. For the section in Fig. 5-40 obtain a plot of $I_E d/h$ as a function of b/d. How may this plot be used as an aid to design?

20. Obtain the uplift pressure on the base of the weir in Fig. 5-30 if the section also contains two symmetrical sheetpiles, as in Fig. 5-32, of lengths (a) $s/b = \frac{1}{4}$, and (b) $s/b = \frac{1}{2}$.

21. If $s = 0$, demonstrate that Eq. (14), Sec. 5-8, reduces to Eq. (9), Sec. 5-7.

22. For the structure shown in Fig. 5-41, obtain the factor of safety with respect to piping and determine soil type for which safety would be assured (a) by exit-gradient criterion, and (b) by creep theory.

23. Obtain the uplift force for the section in Fig. 5-41.

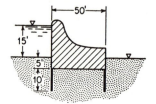

Fig. 5-41

24. In Example 5-3, find the depth of embedment of pilings to increase the factor of safety with respect to piping to 3.0.

25. Demonstrate that the exit gradient at point O of Fig. 5-34 is the minimum exit gradient along the bottom of excavation.

26. The section shown in Fig. 5-42 rests on the surface of an infinite depth of porous media. From first fundamentals derive an expression for (a) the exit gradient and (b) the reduced quantity of seepage, taking any streamline at depth T as an impervious lower boundary.

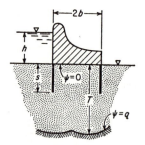

Fig. 5-42

6

Approximate Methods of Solution for Confined Flow Problems

In Chap. 5 we investigated with some rigor the seepage characteristics of various confined flow systems. Although closed-form solutions exist for special structures with two, three, and even four sheetpiles [33, 79, 110], the resulting expressions are generally too complicated for engineering use. Theoretically the solution exists for any configuration; however, with each additional sheetpile or alteration in contour of the structure the flow domain adds two or three new vertices to the Schwarz-Christoffel transformation and hence necessitates the evaluation of hyperelliptic integrals [12, 129]. Thus we are led to approximate solutions.

6-1. Graphical Flow Net

We have already encountered the flow net (Sec. 1-12) and recognized its shortcomings. That is, although the flow net provides the solution (assuming artistic adequacy) for a given domain, should we wish to investigate the effects of a range of characteristic dimensions (such as in a design problem) many flow nets would be required. Consider, for example, the section shown in Fig. 6-1, and suppose we wish to determine the influence of the dimensions A, B, and C on the seepage characteristics, all other dimensions being fixed. Taking only three values for each of these would require 27 individual flow nets.

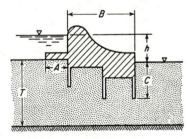

Fig. 6-1

The graphical flow net does serve an important function in providing a check on the adequacy of the ultimate design. In this respect it may be thought of as being analogous to the free-body diagram in mechanics, wherein the physics of the solution can be verified.

141

6-2. Solutions by Analogies: Electrical Analogue

Laplace's equation [Eq. (3), Sec. 1-9], in addition to providing the governing differential equation for the steady-state flow of groundwater, is encountered in many other branches of engineering and physics. Among these are the problems of the steady flow of electricity and heat and various aspects of elastic theory, such as the plane theory of torsion and bending [100]. The reason for this correspondence becomes evident when one considers the nature of the governing laws in these various disciplines; that is, the counterparts of Darcy's law are Fourier's law for heat conduction, Maxwell's law for electrostatics, Ohm's law for current conduction, and Hooke's law and the nature of Airy stress components in plane elastic theory. Although, theoretically, solutions of the Laplace equation for confined seepage problems can be obtained by conducting experiments, with appropriate boundary conditions, on any of the above analogues (cf. Ref. 127), the most productive model experiments have been those of the *electrohydrodynamic type*. The correspondence between the steady-state flow of water through porous media and the steady flow of electric current in a conductor is presented in Table 6-1.

Table 6-1. Correspondence between Seepage and Flow of Electric Current

Steady-state seepage	*Electric current*
Total head h	Voltage V
Coefficient of permeability k	Conductivity σ
Discharge velocity v	Current I
Darcy's law: $v = -k$ grad h	Ohm's law: $I = -\sigma$ grad V
$\nabla^2 h = 0$	$\nabla^2 V = 0$
Equipotential lines: h = constant	Equipotential lines: V = constant
Impervious boundary: $\partial h/\partial n = 0$	Insulated boundary: $\partial V/\partial n = 0$

From Table 6-1 it is evident that the formal analogy between confined seepage flow and current flow is perfect. Thus, to obtain the pattern of equipotential lines for a seepage problem, the flow domain can be transferred into an electrical conductor of similar geometrical form (Fig. 6-2).

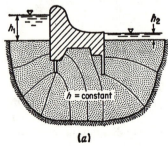

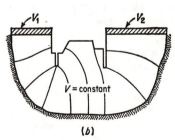

(a)   (b)

Fig. 6-2

The electrical analogue method was first proposed by Pavlovsky in 1918 [109]. After Pavlovsky, experiments of this type were conducted by many others (cf. Refs. 5, 54, 69, 120, 131, 159, 160, 161).* Although the later electrical models are somewhat more elaborate in principle, they differ but little from Pavlovsky's original apparatus, which is shown schematically in Fig. 6-3. Here A represents the flow domain shaped from a high-resistance material. Various metal sheets (monel), heavy paper coated with graphite, special commercial papers (*Teledeltos*), and even liquids (dilute copper-sulfate solutions) and gelatins (*Opsal*) have been used successfully. B is the inflow face at potential V_1, and C is the outflow surface at potential V_2. (The author has had considerable

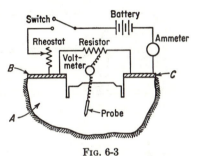

FIG. 6-3

success, for classroom demonstration, using a water-saturated desk blotter with the inflow and outflow faces painted with silver ink, a 6-volt dry-cell battery, a voltmeter, and a paper clip for a probe.) The methodology consists of obtaining (with the probe) the locus of the lines of equal voltage drop which, as noted in Table 6-1, correspond to the location of equipotential lines for a confined flow domain of similar configuration and boundary conditions.

Finally it should be noted that, whereas an electrical analogue simply furnishes the equivalent of a flow net (the streamlines can be sketched in rather easily once the equipotential lines are known), it is subject to the limitations of the flow net (number of experiments required, etc.). An additional limitation on the electrical model is that the electric potential is unaffected by gravity; hence the method requires a priori that the flow system be confined. Also, the finite dimensions of the conducting material severely limit the utility of the method when investigating large spatial problems.

6-3. The Flow Tank

Another experimental tool which has provided some insight into seepage problems is the well-known glass-walled flow tank. In principle, flow-tank experiments represent small-scale reproductions of large-scale flow systems. In essence, one constructs a scale model (generally sand) of the prototype in a tank equipped with a glass front and allows the

---

* Some success has been recorded using the electrical analogue for unconfined flow systems where the conducting material is cut to the shape of the free surface a priori. A method used by De la Marre [27] employed an electrical relaxation procedure.

passage of water (Fig. 6-4). When steady-state flow is reached, a dye is introduced at various points along the upstream boundary close to the glass wall to form the traces of the streamlines. In some models piezometer tubes are also introduced to furnish pressure heads.

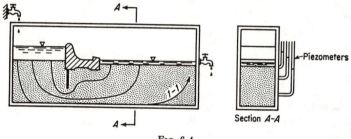

Fig. 6-4

Studies of resulting flow patterns by Schaffernak [126], Dachler [22], Pavlovsky [110], and others have demonstrated conclusively the validity of the fundamental assumptions of groundwater theory. On the basis of his flow-tank experiments Schaffernak verified the generalization of Darcy's law for layered systems and confirmed the action of streamlines at the boundaries of soils with different coefficients of permeabilities (see Sec. 1-14).

Although flow-tank experiments have provided basic information for complicated flow systems, it must be recognized that the very nature of the tank imposes unnatural boundaries to the flow domain. For exam-

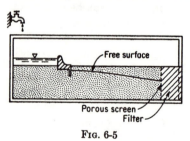

Fig. 6-5

ple, in Fig. 6-4, the downstream side of the tank forces streamlines, such as 1-1, to intersect the tail-water boundary rather than seek more natural subsurface drainage outlets. In such situations the experiment itself induces confinement and prevents any possibilities of the development of a free surface. A flow tank such as illustrated in Fig. 6-5 will demonstrate an entirely different pattern of flow. Another drawback of this analogy is the development of capillary heads in the sand, which tend to overshadow gravity effects in the small model but which would be inconsequential in the much larger prototype.

6-4. Viscous Flow Models: Hele-Shaw Model

For the steady-state flow of viscous, incompressible fluids (at small Reynolds numbers), the Navier-Stokes equations of motion, the most

general equations governing fluid flow, reduce in form to generalized statements of Darcy's law. Recognizing this relationship, Hele-Shaw [55, 56] in 1897 devised an apparatus (which bears his name) whereby two-dimensional groundwater flow could be investigated experimentally for structures with complex boundaries. Essentially, the model consists

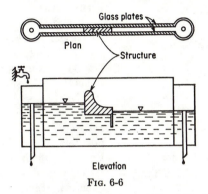

Fɪɢ. 6-6

of two closely spaced glass plates containing completely the shape of the structure to be investigated (Fig. 6-6). A viscous fluid such as glycerine [15, 57] is then allowed to flow between inlet and outlet levels until steady-state flow is reached. Then, by injecting colored dyes along the upstream edge, the patterns of streamlines can be observed.*

The Navier-Stokes equations of motion may be expressed as [141]

$$X - \frac{1}{\rho}\frac{\partial p}{\partial x} + \frac{\nu}{3}\frac{\partial}{\partial x}\left(\frac{\partial u}{\partial x} + \frac{\partial v}{\partial y} + \frac{\partial w}{\partial z}\right) + \nu\nabla^2 u = \frac{Du}{Dt}$$

$$Y - \frac{1}{\rho}\frac{\partial p}{\partial y} + \frac{\nu}{3}\frac{\partial}{\partial y}\left(\frac{\partial u}{\partial x} + \frac{\partial v}{\partial y} + \frac{\partial w}{\partial z}\right) + \nu\nabla^2 v = \frac{Dv}{Dt} \qquad (1)$$

$$Z - \frac{1}{\rho}\frac{\partial p}{\partial z} + \frac{\nu}{3}\frac{\partial}{\partial z}\left(\frac{\partial u}{\partial x} + \frac{\partial v}{\partial y} + \frac{\partial w}{\partial z}\right) + \nu\nabla^2 w = \frac{Dw}{Dt}$$

where x, y, z = spatial coordinates
u, v, w = velocity components
X, Y, Z = body forces in x, y, z directions, respectively
$\rho = \gamma/g$, mass density of fluid
$\nu = \mu/\rho$ kinematic viscosity
$$\nabla^2 = \frac{\partial^2}{\partial x^2} + \frac{\partial^2}{\partial y^2} + \frac{\partial^2}{\partial z^2}$$
$$\frac{D}{Dt} = \frac{\partial}{\partial t} + u\frac{\partial}{\partial x} + v\frac{\partial}{\partial y} + w\frac{\partial}{\partial z}$$

* For a more detailed development of the history of the apparatus and the results of some recent experiments see dissertation by D. K. Todd [146].

Letting $\nu = 0$, we see that the Navier-Stokes equations reduce to the Euler equations, Eq. (2), Sec. 1-8.

If the distance $2a$ between the plates is small, the flow becomes two-dimensional ($w = 0$). In rectangular coordinates the x axis will be

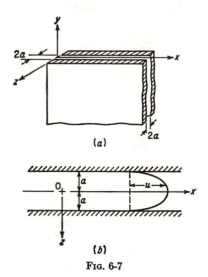

chosen as horizontal and midway between the plates, the y axis vertical, and the z axis perpendicular to the plates (Fig. 6-7). Because the only body forces are due to gravity, assuming a gravity potential of $\Omega = gH$,* where H is the elevation head ($H = h - p/g\rho$), we have $X = -\partial(gH)/\partial x$, $Y = -\partial(gH)/\partial y$, and $Z = -\partial(gH)/\partial z$. For incompressible fluids, the third term in Eqs. (1), which represents the rate of volume dilation of the fluid, vanishes. Since the velocity at the glass plates must be zero (Fig. 6-7b), the change in the velocity components (u and v) with respect to z will be much greater than the velocity changes with respect to the x and

(a)

(b)

Fig. 6-7

y directions. Hence, in comparison to their respective derivatives in the z direction, we may assume $\partial u/\partial x$, $\partial u/\partial y$, $\partial v/\partial x$, $\partial v/\partial y$, and their second derivatives are of negligible order. Now, assuming steady-state conditions and making the stated assumptions, Eqs. (1) reduce to

$$\frac{\partial h}{\partial x} = \nu \frac{\partial^2 u}{\partial z^2}$$
$$\frac{\partial h}{\partial y} = \nu \frac{\partial^2 v}{\partial z^2} \tag{2}$$
$$\frac{\partial h}{\partial z} = 0$$

The third of Eqs. (2) shows only a hydrostatic variation in pressure in the z-direction; that is, the total head at any point within the flow domain will depend upon the x and y coordinates only. Hence, the first two equations can be integrated with respect to z to yield

$$z \frac{\partial h}{\partial x} = \nu \frac{\partial u}{\partial z} \qquad z \frac{\partial h}{\partial y} = \nu \frac{\partial v}{\partial z}$$

* See Prob. 10, Chap. 1.

where the constants of integration are zero, as (from symmetry)

$$\frac{\partial u}{\partial z} = \frac{\partial v}{\partial z} = 0$$

at $z = 0$. Integrating once more with respect to z and noting that $u = v = 0$ at $z = \pm a$, we obtain for the velocity components

$$u = -\frac{z^2 - a^2}{2\nu}\frac{\partial h}{\partial x}$$
$$v = -\frac{z^2 - a^2}{2\nu}\frac{\partial h}{\partial y} \tag{3}$$

which demonstrate a parabolic velocity distribution (Fig. 6-7b). In the one-dimensional case, such flow is called *plane Poiseuille flow.** Finally, for the mean velocity, from Eqs. (3) we obtain

$$\bar{u} = -\frac{a^2 g}{3\nu}\frac{\partial h}{\partial x} = -m\frac{\partial h}{\partial x}$$
$$\bar{v} = -\frac{a^2 g}{3\nu}\frac{\partial h}{\partial y} = -m\frac{\partial h}{\partial y} \tag{4}$$

Calling m the *coefficient of permeability of the channel* between the glass plates, we see that the analogy with Darcy's law is complete.

Taking Darcy's law as given in Eq. (3), Sec. 1-6, and forming a ratio with the results given above, we obtain for the velocity ratio between the model and prototype, for the same gradients,

$$V_r = \frac{V_m}{V_p} = \frac{a^2 \rho_r}{3k_0 \mu_r} \tag{5}$$

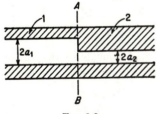

FIG. 6-8

where the subscript m applies to the model, p to the prototype, and r to the ratio of model to prototype, and k_0 is the physical permeability.

The Hele-Shaw model can also be used for studies of nonhomogeneous conditions. With reference to Fig. 6-8 it can easily be shown (by a procedure analogous to that used in Sec. 1-14, Subsection 3) that the effective coefficients of permeability in each of the regions are related by [116]

$$\frac{k_1}{k_2} = \frac{a^2_1}{a^2_2} \tag{6}$$

---

* For an interesting development of the flow characteristics through porous media based on Poiseuille flow, see Taylor [142].

6-5. Relaxation Method

A numerical procedure which has been used with some success [98] to obtain approximate solutions of complex flow problems is the so-called *relaxation process* [139] based upon the calculus of finite differences. Essentially the procedure consists of reducing a partial differential equation in the vicinity of a point into an algebraic difference equation.

Referring to the Laplace equation for the velocity potential

$$\frac{\partial^2 \phi}{\partial x^2} + \frac{\partial^2 \phi}{\partial y^2} = 0 \tag{1}$$

we find for the square mesh in Fig. 6-9a, assuming that the distance a is

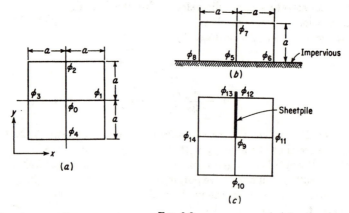

Fig. 6-9

small enough so that a linear variation between adjacent ϕ's is not greatly in error,

$$\frac{\partial^2 \phi}{\partial x^2} \approx \frac{(\phi_1 - \phi_0)/a - (\phi_0 - \phi_3)/a}{a} = \frac{\phi_1 + \phi_3 - 2\phi_0}{a^2}$$

$$\frac{\partial^2 \phi}{\partial y^2} \approx \frac{\phi_2 + \phi_4 - 2\phi_0}{a^2}$$

whence we obtain

$$4\phi_0 \approx \phi_1 + \phi_2 + \phi_3 + \phi_4 \tag{2}$$

Equation (2) is said to satisfy Laplace's equation at an *interior node* and is commonly represented as shown in Fig. 6-10a. Although Eq. (2) is an approximation, equality is approached as the size of the grid decreases. The characteristic nodes at various boundary points can be determined in like manner; however, it is generally more convenient to determine

h (without subscript) is the total head loss through the section [6]-12). By similar reasoning we find that the head loss in the mth [e]nt can be calculated from

$$h_m = \frac{h\Phi_m}{\Sigma\Phi} \qquad (3)$$

[t]he head loss for any fragment has been determined, the pressure [sol]ution on the base of the structure and the exit gradient can be [o]btained. Thus, the primary task is to implement this method by [establi]shing a catalogue of typical form factors. Following Pavlovsky's [procedu]re, the various form factors will be divided into types, and the [charact]eristics of each type will be studied. Finally, the results will be [summa]rized in tabular form for easy reference.

[Type] I (Fig. 6-13). The fragment of type I is a region of parallel [horizon]tal flow between impervious boundaries. From Darcy's law, we have simply $q = kah/L$, and hence the form factor is

$$\Phi = \frac{L}{a} \qquad (4a)$$

For an elemental section (Fig. 6-13b),

$$d\Phi = \frac{dx}{y} \qquad (4b)$$

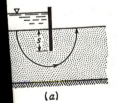

(b)

FIG. 6-13

Obviously, the pressure distribution for the type I fragment is linear.

[Type] II (Fig. 6-14). From the results of Prob. 12, Chap. 5 (also as a [special c]ase of Eqs. 14, Sec. 5-5, with $b = 0$), we find that the discharge

(a)

(b)

(c)

FIG. 6-14

[a]round a single sheetpile of embedment s in a layer of thickness [T (Fig. 6]-14a) is

$$q = \frac{khK'}{2K}$$

[with mod]ulus $m = \sin(\pi s/2T)$. Considering the type II fragment as [one of] the sections in Fig. 6-14b or c (b is an entrance condition, c an [exit cond]ition), we have for these cases $q = kh_m K'/K$, where h_m is taken as the [head loss] through the fragment. The modulus as given above can

them directly from Darcy's law. For example, for the square mesh in Fig 6-9a, from Darcy's law

$$q_{3\to0} = va = \frac{\partial\phi}{\partial x} a \approx \frac{\phi_3 - \phi_0}{a} a = \phi_3 - \phi_0$$

$$q_{2\to0} \approx \phi_2 - \phi_0$$

$$q_{0\to4} \approx \phi_0 - \phi_4$$

$$q_{0\to1} \approx \phi_0 - \phi_1$$

Since $q_{in} - q_{out} = 0$, we obtain indentically Eq. (2).

Considering now the node at an impervious boundary (Fig. 6-9b),

$$q_{8\to5} \approx \frac{\phi_8 - \phi_5}{a}\frac{a}{2}$$

$$q_{5\to6} \approx \frac{\phi_5 - \phi_6}{a}\frac{a}{2}$$

$$q_{5\to7} \approx \phi_5 - \phi_7$$

and hence $\qquad \phi_7 + \frac{\phi_8}{2} + \frac{\phi_6}{2} \approx 2\phi_5 \qquad (3)$

The node representing Eq. (3) is shown in Fig. 6-10b. We note that the form of the node is directly related to the width of the *channel* between

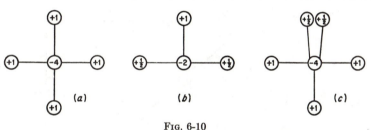

(a)

(b)

(c)

FIG. 6-10

the points under consideration; that is, in Fig. 6-9b between ϕ_6, ϕ_8, and ϕ_5 there is only one-half the width available for flow that there is between ϕ_7 and ϕ_5. Hence the form of the node can be obtained by inspection. For example, the node for the bottom of a piling such as in Fig. 6-9c is as shown in Fig. 6-10c. By a similar procedure the form of the nodes can be obtained for any boundary within the flow domain.

Once the various nodal conditions have been established for a particular flow domain, the determination of the values of ϕ at any grid crossing (lattice point) can be obtained by a relaxation procedure. Generally the procedure involves an evaluation of the residual at a point and a systematic refinement of this residual throughout the entire net. Although several methods are available for effecting the reduction of the residual, the procedure described below is recommended for its simplicity and ease of visual checking. To illustrate the method of solution, we shall investigate the example shown in Fig. 6-11a. Because of symmetry only half of the domain need be considered.

1. On a scale drawing of the section construct a rough flow net (shown dotted in Fig. 6-11a) and assign corresponding values of ϕ to lattice points. The first approximation of the ϕ values is shown on the figure.

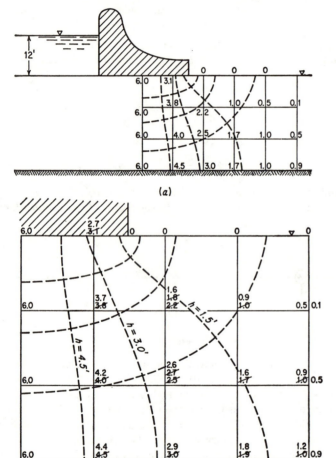

(a)

(b)

FIG. 6-11

2. Above each lattice point enter its value based upon the characteristic of the node (Fig. 6-10). This step is called *relaxing the node*, and although the nodes can be relaxed in any order a systematic procedure is desirable. Thus in Fig. 6-11b the sequence of operations follows from upper left to lower right.

3. The procedure is then repeated until the differen obtained within desired limits. In this example only necessary because of the refinement of the original rougher net been drawn originally (or none at all) th still yield the same solution; however, many more been required.

4. Finally the equipotential lines are drawn, bas ϕ values, and the flow pattern is established (dotted

Although the above procedure provides a positive mining the pattern of flow to any degree of approx emphasized that a solution can be obtained only for a Thus the method is subject to all the limitations pr the graphical flow net, except that the procedure is

For additional information on relaxation method problems including unconfined flow conditions, se and 139. For other numerical solutions of Laplace 38 and 97.

6-6. Method of Fragments

An approximate analytical method of solution for any confined flow system of finite depth, directly applicable to design, was furnished by Pavlovsky[110] in 1935. The fundamental assumption of this method, called the *method of fragments*, is that equipotential lines at various critical parts of the flow region can be approximat lines (as, for example, the dotted lines in Fig. 6-12 into sections or fragments (Fig. 6-12 is shown divid Suppose, now, that we can compute the discharge

$$q = \frac{kh_m}{\Phi_m} \qquad m = 1, 2, \ldots$$

where h_m = head loss through fragment
Φ_m = dimensionless form factor*
Then, since the discharge through all fragments

$$q = \frac{kh_1}{\Phi_1} = \frac{kh_2}{\Phi_2} = \frac{kh_m}{\Phi_m} = \cdots$$

and

$$q = k \frac{\Sigma h_m}{\Sigma \Phi} = \frac{kh}{\sum\limits_{m=1}^{n} \Phi_m}$$

* Compare with N_e/N_f in Eq. (4), Sec. 1-12.

where
(Fig.
fragm

Once
distrib
easily
establi
proced
charac
summa

Type
horizon

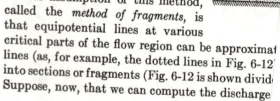

Type
special

for flow
T (Fig.

with mo
either of
exit condi
head loss

be obtained directly from Fig. 5-15.   Hence the form factor is

$$\Phi = \frac{K}{K'} \qquad m = \sin \frac{\pi s}{2T} \tag{5a}$$

The exit gradient (at point E in Fig. 6-14c) will be [Eq. (16a), Sec. 5-5]

$$I_E = \frac{h_m \pi}{2KTm} \qquad m = \sin \frac{\pi s}{2T} \tag{5b}$$

where h is again the head loss through the fragment.   The solution of Eq. (5b) can be obtained directly from Fig. 5-22 (taking h in the figure as twice that in the fragment).

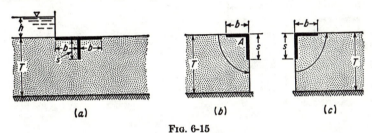

Fig. 6-15

Type III (Fig. 6-15).   From Eq. (14c), Sec. 5-5, the discharge for Fig. 6-15a is $q = khK'/2K$, where the modulus is given by Eq. (14b), Sec. 5-5.   Hence for either of the fragments of Fig. 6-15b or c

$$\Phi = \frac{K}{K'} \tag{6a}$$

where the modulus

$$m = \cos \frac{\pi s}{2T} \sqrt{\tanh^2 \frac{\pi b}{2T} + \tan^2 \frac{\pi s}{2T}} \tag{6b}$$

can be obtained directly from Fig. 5-15.

Type IV (Fig. 6-16).   Pavlovsky considers the sections shown in Fig. 6-16a as his type IV fragments.   The exact solution for this fragment is

$$\Phi = \frac{K'}{K}$$

with the modulus

$$m = \lambda \operatorname{sn} \left(\frac{a}{T} \Lambda, \lambda \right)$$

where Λ = complete elliptic integral of first kind of modulus λ

   Λ' = complete elliptic integral of first kind of complementary modulus λ'

$$\frac{\Lambda}{\Lambda'} = \frac{T}{b}$$

The method of solution will now be demonstrated by an example.

Example 6-1. Determine the form factor for Fig. 6-16a if $T = 12$ ft, $b = 8$ ft, and $a = 4$ ft.

For this case the ratio $\Lambda/\Lambda' = T/b = 12/8 = 1.5$. Hence from Table B-1, we find $\lambda = 0.93$ and $\Lambda = 2.441$. With known values substituted, the modulus is $m = 0.93$ sn $(2.441/3, 0.93) = 0.634$ and the form factor is $\Phi = 1.10$.

To simplify the solution, Pavlovsky noted from his electrical analogues that the quantity of seepage above the streamline AB of Fig. 6-16b was of small order and could be neglected. Hence he divided the flow region into two parts, labeled active and passive in Fig. 6-16c, with the dividing

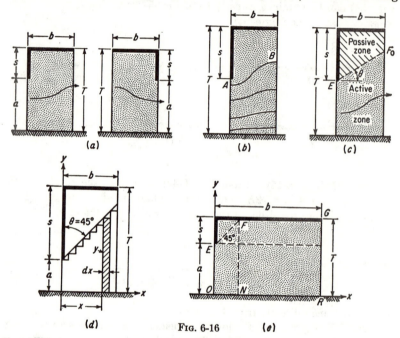

(a) (b) (c)

(d) Fig. 6-16 (e)

line EF_0 at an angle θ. On the basis of his analogue studies, Pavlovsky assumed $\theta = 45°$. With this assumption, two conditions need to be considered for type IV fragments, depending on the ratio of b to s.

1. $b \leqq s$. For this case (Fig. 6-16d), following Pavlovsky, we shall consider the active zone to be composed of elements of type I fragments of width dx. Hence

$$\Phi = \int_0^b \frac{dx}{y} = \int_0^b \frac{dx}{a + x}$$

and the form factor is

$$\Phi = \ln\left(1 + \frac{b}{a}\right) \tag{7a}$$

Substituting the values of Example 6-1, we find that the approximate solution also gives $\Phi = 1.10$.

2. $b \geqq s$. For this case (Fig. 6-16e)

$$\Phi = \int_0^s \frac{dx}{a+x} + \int_s^b \frac{dx}{T}$$

and the form factor is

$$\Phi = \ln\left(1 + \frac{s}{a}\right) + \frac{b-s}{T} \tag{7b}$$

Type V (Fig. 6-17).   We see from Fig. 6-17 that the form factor for the
type V fragment is twice that of the type IV
fragment; hence for $L \leqq 2s$,

$$\Phi = 2 \ln\left(1 + \frac{L}{2a}\right) \tag{8a}$$

and for $L \geqq 2s$

$$\Phi = 2 \ln\left(1 + \frac{s}{a}\right) + \frac{L - 2s}{T} \tag{8b}$$

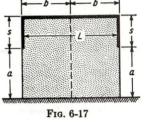

Fig. 6-17

Type VI (Fig. 6-18).   Using the same approximations as for the
type IV fragments, we see that two cases are to be considered.

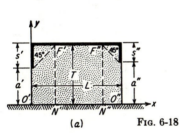

(a)   Fig. 6-18   (b)

1. $L \geqq s' + s''$. Noting in this case (Fig. 6-18a) that

$$\Phi = \int_0^{s'} \frac{dx}{a'+x} + \int_{s'}^{L-s''} \frac{dx}{T} + \int_{L-s''}^{L} \frac{dx}{a''+L-x}$$

we obtain

$$\Phi = \ln\left[\left(1 + \frac{s'}{a'}\right)\left(1 + \frac{s''}{a''}\right)\right] + \frac{L - (s' + s'')}{T} \tag{9a}$$

2. $L \leqq s' + s''$. For this case (Fig. 6-18b), we have

$$\Phi = \int_0^{b_1} \frac{dx}{a'+x} + \int_{b_1}^{L} \frac{dx}{a''+L-x}$$

hence

$$\Phi = \ln\left[\left(1 + \frac{b'}{a'}\right)\left(1 + \frac{b''}{a''}\right)\right] \tag{9b}$$

where

$$b' = \frac{L + (s' - s'')}{2}$$

$$b'' = \frac{L - (s' - s'')}{2}$$

Table 6-2. Summary of Fragment Types and Form Factors

Fragment type	Illustration	Φ—Form factor (h is head loss through fragment)
I	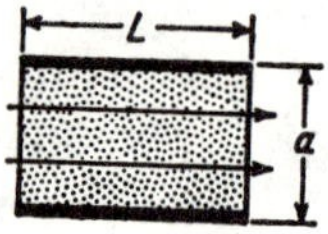	$\Phi = \dfrac{L}{a}$
II	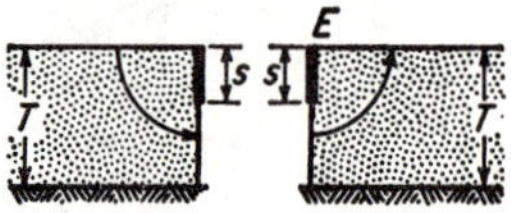	$\Phi = \dfrac{K}{K'};\ m = \sin\dfrac{\pi s}{2T}$ $I_E = \dfrac{h_m \pi}{2KT_m}$ (See Fig. 5-22.)
III	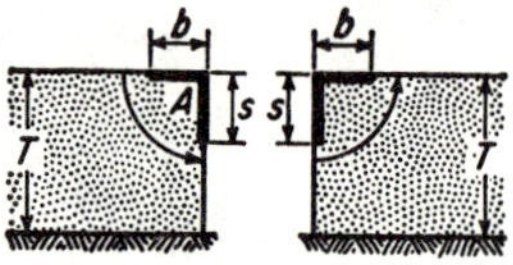	$\Phi = \dfrac{K}{K'}$ $m = \cos\dfrac{\pi s}{2T}\sqrt{\tanh^2\dfrac{\pi b}{2T} + \tan^2\dfrac{\pi s}{2T}}$ (See Fig. 5-15.)
IV	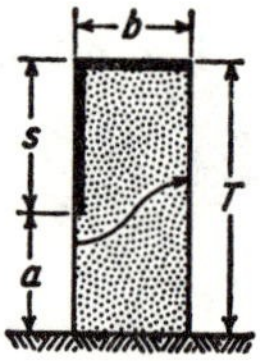 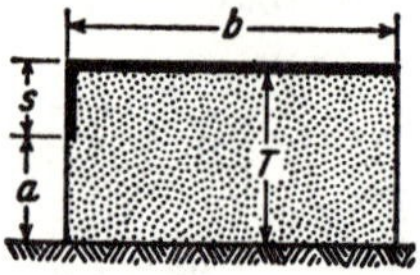	*Exact solution* (see Example 6-1): $\dfrac{\Lambda}{\Lambda'} = \dfrac{T}{b}$; modulus $= \lambda$ $\Phi = \dfrac{K'(m)}{K(m)};\ m = \lambda\ \text{sn}\left(\dfrac{a}{T}\Lambda, \lambda\right)$ *Approximate solution:* $S \geqq b$: $\qquad \Phi = \ln\left(1 + \dfrac{b}{a}\right)$ $b \geqq S$: $\qquad \Phi = \ln\left(1 + \dfrac{s}{a}\right) + \dfrac{b - s}{T}$

Table 6-2.—*(Continued)*

Fragment type	Illustration	Φ—Form factor (h is head loss through fragment)
V		$L \leqq 2s$: $$\Phi = 2 \ln\left(1 + \frac{L}{2a}\right)$$ $L \geqq 2s$: $$\Phi = 2 \ln\left(1 + \frac{s}{a}\right) + \frac{L - 2s}{T}$$
VI		$L > s' + s''$: $$\Phi = \ln\left[\left(1 + \frac{s'}{a'}\right)\left(1 + \frac{s''}{a''}\right)\right]$$ $$\quad + \frac{L - (s' + s'')}{T}$$ $L = s' + s''$: $$\Phi = \ln\left[\left(1 + \frac{s'}{a'}\right)\left(1 + \frac{s''}{a''}\right)\right]$$ $L < s' + s''$: $$\Phi = \ln\left[\left(1 + \frac{b'}{a'}\right)\left(1 + \frac{b''}{a''}\right)\right]$$ where $$b' = \frac{L + (s' - s'')}{2}$$ $$b'' = \frac{L - (s' - s'')}{2}$$

The various fragment types and pertinent relationships are presented in Table 6-2 for easy reference.

To determine the pressure distribution on the base of a structure (such as that along $C'CC''$ in Fig. 6-19), we shall assume that the head loss within the fragment is linearly distributed along the impervious boundary. Thus, in Fig. 6-19, if h_m is the head loss within the fragment, the rate of loss along $E'C'CC''E''$ will be

$$R = \frac{h_m}{L + s' + s''} \tag{10}$$

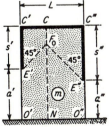

Fig. 6-19

Once the total head is known at any point, the pressure can easily be determined.

Example 6-2. For the section shown in Fig. 6-20a, estimate (a) the reduced discharge, (b) the uplift pressure on the base of the structure, and (c) the exit gradient.

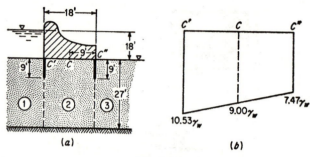

Fig. 6-20

The division of fragments is shown on the figure. Regions 1 and 3 are both type II fragments, and the middle section is of type V with $L = 2s$.

For regions 1 and 3, we have $m = \sin{(\pi s/2T)} = 0.5$; hence (from Table B-1) we obtain $\Phi_1 = \Phi_3 = K/K' = 0.782$.

For region 2, since $L = 2s$, $\Phi_2 = 2 \ln{(1 + 18/36)} = 0.811$.

Thus the sum of the form factors is

$$\Sigma\Phi = 0.782 + 0.811 + 0.782 = 2.375$$

and the reduced quantity of seepage [Eq. (2)] is

$$\frac{q}{k} = \frac{18}{2.375} = 7.6 \text{ ft}$$

For the head loss in each of the sections, from Eq. (3) we find

$$h_1 = h_3 = \frac{0.782}{2.375} 18 = 5.93 \text{ ft}$$
$$h_2 = 6.14 \text{ ft}$$

Hence the head-loss rate [Eq. (10)] is

$$R = \frac{6.14}{36} = 17 \text{ per cent}$$

and the pressure distribution along $C'CC''$ is as shown in Fig. 6-20b.

For the exit gradient [Eq. (5b)], from Fig. 5-22 with $s/T = \frac{1}{3}$, we find $I_E s/h = 0.31$, where h as given in this figure is twice the head loss in the type II fragment. Hence

$$I_E = \frac{0.31 \times 2 \times 5.93}{9} = 0.41$$

From Eq. (14), Sec. 5-8 (for an infinite depth of porous media), the exit gradient is found to be 0.43.

6-7. Flow in Layered Systems

Closed-form solutions for the seepage characteristics of even simple structures founded in layered media offer considerable mathematical difficulty. On the basis of her closed-form solutions for the sections shown in Fig. 6-21, in 1941 Polubarinova-Kochina [114] developed an

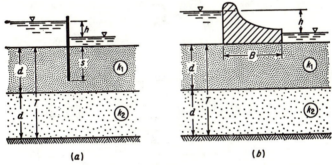

(a)   (b)

Fig. 6-21

approximate procedure whereby the seepage characteristics of structures founded in layered systems can be obtained simply and with a great degree of reliability. The procedure will be illustrated for the sections shown in Fig. 6-21 for which the exact solutions are known. In these figures the flow media consist of two horizontal layers of equal thickness d underlain by an impervious base. The coefficient of permeability of the upper layer is k_1, and of the lower layer k_2, and the coefficients of permeability are related to the dimensionless parameter ε by the expression

$$\tan \pi\varepsilon = \sqrt{\frac{k_2}{k_1}} \qquad (1)*$$

Thus, as the ratio of the permeabilities varies from 0 to ∞, ε ranges between 0 and $\frac{1}{2}$.

Let us investigate the discharge and the exit gradient for the structures shown in Fig. 6-21 for some special values of ε.

* Compare with Sec. 1-14, Subsection 3.

Single Sheetpile Embedded in Two Layers of Equal Thickness and of Different Coefficients of Permeability (Fig. 6-21a)

1. $\varepsilon = 0$. When $\varepsilon = 0$, from Eq. (1) we have $k_2 = 0$, which is equivalent to having the impervious base at depth d. Hence for this case the flow region is reduced to a single homogeneous layer for which the discharge and exit gradient are known. From Eq. (14), Sec. 5-5, taking $b = 0$, we get for the discharge

$$q = \frac{k_1 h K'}{2K} \qquad m = \sin \frac{\pi s}{2d} \qquad (2)$$

From Eqs. (16), Sec. 5-5, the exit gradient is

$$I_E = \frac{h\pi}{4K\,dm} \qquad (3)$$

In the above, $s < d$. If $s \geqq d$, obviously $q = I_E = 0$.

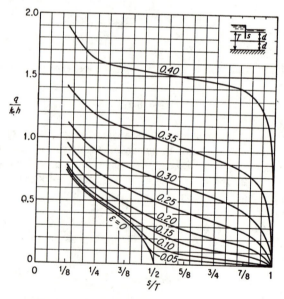

Fig. 6-22. (*After Polubarinova-Kochina* [116].)

The curve for q/k_1h as a function of s/T for $\varepsilon = 0$, where T is the thickness of both layers, is given in Fig. 6-22. A crossplot is shown for various s/T ratios (at $\varepsilon = 0$) in Fig. 6-23. Similar plots for $I_E T/h$ are presented in Fig. 6-24.

2. $\varepsilon = \frac{1}{4}$. When $\varepsilon = \frac{1}{4}$, $k_2 = k_1$, and the system is reduced to a single homogeneous layer of thickness $2d$, for which Eqs. (2) and (3) are applicable (taking d in these expressions as $2d$). The corresponding plots for $\varepsilon = \frac{1}{4}$ are given in Figs. 6-22 to 6-24.

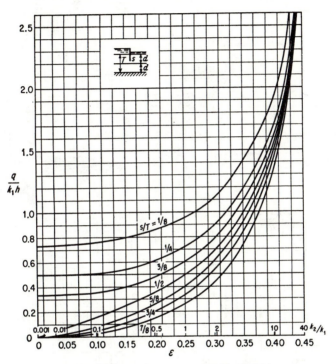

Fig. 6-23. (*After Polubarinova-Kochina* [116].)

3. $\varepsilon = \frac{1}{2}$. When $\varepsilon = \frac{1}{2}$, $k_2 = \infty$, and there is no resistance to flow in the bottom layer. Hence the discharge through the section is infinite, i.e., $q = k_1 h K'/2K = \infty$. Now, as K'/K must be infinite for this case, for values of ε in the vicinity of $\frac{1}{2}$, Polubarinova-Kochina recommends the approximate equality of $K'/K = \tan \varepsilon\pi$. Thus, for values of ε close to $\frac{1}{2}$, the discharge can be determined by

$$q = \frac{k_1 h}{2} \sqrt{\frac{k_2}{k_1}} \tag{4}$$

The closed-form solution yields for this case

$$\lim_{\varepsilon \to \frac{1}{2}} \left(\frac{q/k_1 h}{\tan \varepsilon\pi} \right) = \frac{1}{2}$$

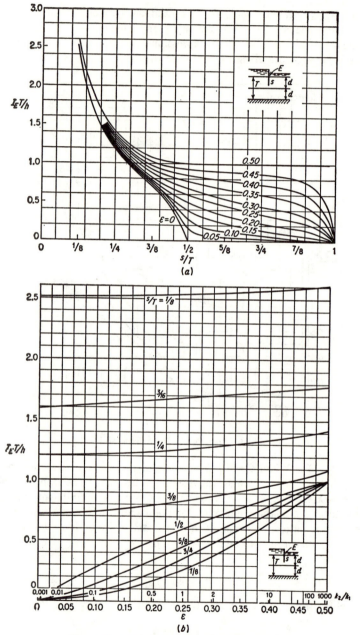

FIG. 6-24. (*After Polubarinova-Kochina* [116].)

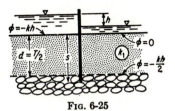

FIG. 6-25

If $s \geq d$, the flow through the upper layer is one-dimensional (Fig. 6-25), and hence the exit gradient is $I_E = h/T$, or

$$I_E \frac{T}{h} = 1 \tag{5}$$

To determine the exit gradient for $s < d$, it is necessary to proceed as in Chap. 5 and find the correspondence between the z plane and the w plane in Fig. 6-26. The mechanics of the solution will be left for the

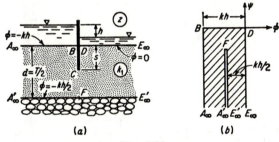

FIG. 6-26

problems.* The complex velocity for this problem is found to be ($T = 2d$)

$$u - iv = -\frac{k_1 h}{T} \frac{\cosh(\pi z/T)}{\sqrt{\sinh^2(\pi z/T) + \sin^2(\pi s/T)}} \tag{6}$$

and hence ($z = 0$ at the exit point)

$$I_E \frac{T}{h} = \frac{1}{\sin(\pi s/T)} \tag{7}$$

The results given in Eqs. (5) and (7) are shown in Fig. 6-24.

Flat-bottom Structure Resting on the Surface of Two Layers of Equal Thickness and of Different Coefficients of Permeability

For this section (Fig. 6-21b) the exit gradient will be unbounded for all cases; hence we need only be concerned with the question of the discharge as a function of the permeability ratios. For both $\epsilon = 0$ and $\epsilon = \frac{1}{4}$ we

* See Girinsky [43].

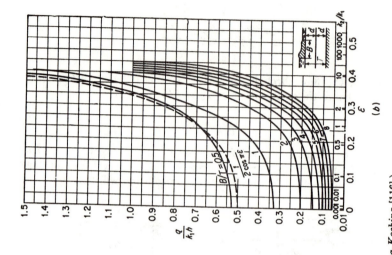

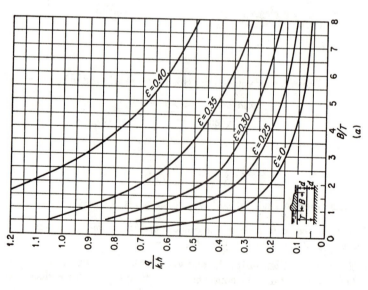

Fig. 6-27. (*After Polubarinova-Kochina [116].*)

find that the discharge [Eqs. (14), Sec. 5-5, with $s = 0$] is given by

$$q = \frac{k_1 h K'}{2K} \qquad m = \tanh \frac{\pi B}{2d} \tag{8}$$

where d is to be taken as $2d$ (or T) for $\varepsilon = \frac{1}{4}$. Figure 6-27 shows the variation of this function for values of B/T where T is the thickness of both layers. For values of ε in the vicinity of $\frac{1}{2}$, Polubarinova-Kochina again recommends the use of the approximate relationship given in Eq. (4).

The foregoing demonstrates that for the special values of $\varepsilon = 0$, $\varepsilon = \frac{1}{4}$, and $\varepsilon = \frac{1}{2}$, the problem of determining the flow characteristics for structures founded in two layers of different permeabilities can be reduced to one for a single homogeneous layer. This provides the essentials of Polubarinova-Kochina's approximate method wherein exact solutions are obtained and plotted for values of $\varepsilon = 0$ and $\varepsilon = \frac{1}{4}$ and in the near vicinity of $\varepsilon = \frac{1}{2}$, and smooth curves are drawn between these known points (such as in Figs. 6-23, 6-24b, and 6-27b), from which intermediate values can be obtained by interpolation.

A simplification in the above method when the discharge becomes infinite can be effected by plotting the inverse of the ordinate scale for the discharge; that is, plot the curves for $k_1 h/q$ rather than $q/k_1 h$ versus ε, as was done in Fig. 6-27b. This obviates the difficulty at $\varepsilon = \frac{1}{2}$, which now becomes a known point $(k_1 h/q = 0)$. For portions of curves such as $s/T = \frac{1}{2}$ in Fig. 6-23, where $q/k_1 h \to 0$ as $\varepsilon \to 0$, part of the curves can be obtained from plots using $k_1 h/q$ (at $\varepsilon = \frac{1}{4}$ and $\varepsilon = \frac{1}{2}$), and part using $q/k_1 h$ (for $\varepsilon = 0$ and $\varepsilon = \frac{1}{4}$).

The above method can be used as well for layers of unequal thickness. For example, in Fig. 6-28 for $\varepsilon = 0$, the equivalent depth of the flow domain is d, whereas at $\varepsilon = \frac{1}{4}$ the depth is $3d$.

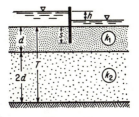

FIG. 6-28

Finally, it should be noted that this procedure can be combined with Pavlovsky's method of fragments to yield approximate solutions for even the most complicated structures.

PROBLEMS

1. Draw a graphical flow net for the flow domain shown in Fig. 6-29, and determine the discharge quantity.

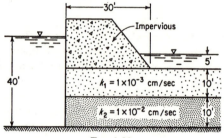

FIG. 6-29

2. Demonstrate the correspondence, as in Table 6-1, between groundwater flow and (a) the flow of heat and (b) elastic theory.

3. Demonstrate that the third term in each of Eqs. (1), Sec. 6-4, is zero for incompressible flow.

4. Glycerine at 80°F is to be used in a Hele-Shaw model experiment. The coefficient of permeability in the prototype is 0.01 cm/sec. What channel width would give a 1:1 correspondence between model and prototype? What would the required spacing be for distilled water at the same temperature?

5. Prove Eq. (3), Sec. 6-4.

6. Using the relaxation method, obtain the flow net for the section shown in Fig. 6-30. Locate the equipotential lines $h/3$, $h/5$, and $2h/3$.

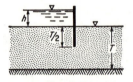

FIG. 6-30

7. Demonstrate that the node for Fig. 6-9c is as shown in Fig. 6-10c.

8. Find the interior node at the boundary between the two layers in Fig. 6-29.

9. For the section shown in Fig. 6-31, obtain plots of (a) the reduced quantity of seepage and (b) the exit gradient. Assume $0 \leq A \leq 75$ ft.

10. If $A = 25$ ft in Fig. 6-31, obtain the pressure distribution along the base of the structure.

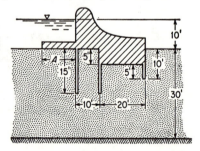

FIG. 6-31

11. Find the exact pressure acting at point C of Example 6-2.

12. Solve Prob. 26, Chap. 5, by the method of fragments.

13. Estimate the discharge through the section in Fig. 6-29.

14. Obtain for the section in Fig. 6-32 the graphical solution for the quantity of seepage (q/k_1h) as a function of k_2/k_1 for $d/2 \leqq B \leqq 3d$ and $s/d = \frac{1}{2}$.

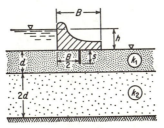

Fig. 6-32

15. In Fig. 6-33, $k_2 = 8k_1 = 0.0016$ cm/sec. Obtain a plot of (a) the discharge and (b) the factor of safety with respect to piping, as a function of the depth of sheet-pile embedment.

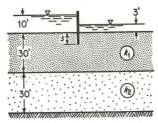

Fig. 6-33

16. Explain the close agreement in Fig. 6-27b between the curve labeled $\frac{1}{2} \cos \pi\varepsilon$ (shown dotted) and the curve for $B/T = 0.5$.

17. For the section shown in Fig. 6-34 estimate (a) the reduced quantity of seepage (q/k_1) for $k_2 = 4k_1$ and $k_2 = k_1/2$, (b) the pressure in the water at point A if $k_2 = k_1/2$, and (c) the exit gradient if $k_2 = 3k_1$.

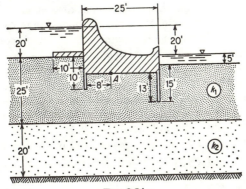

Fig. 6-34

7

Unconfined Flow through Earth Structures on Homogeneous Foundations of Great Depth

7-1. General Discussion

Having considered in some detail the main properties of confined flow, we shall, in the following chapters, turn our attention to the class of problems wherein at least one boundary of the flow domain is a *free surface*. It was demonstrated in previous discussions (cf. Sec. 1-11) that, although a free surface is a streamline along which the pressure is constant, its locus is not known a priori. Thus, the motion of ground-water in any flow system wherein the boundaries of the domain are not rigidly prescribed is said to exhibit *unconfined flow*. In this chapter we shall investigate those flow systems in which the depth to an impervious stratum is sufficiently great to be infinitely extended.* A broader treatment of the problem will be given in Chap. 8.

7-2. Unconfined Flow around Cutoffs

It is imperative at this point that the question of confinement be examined judiciously. The literature is replete with solutions to problems wherein confinement has been specified arbitrarily, much as was done in Chap. 5. In view of the simplifications derivable from this procedure, it has been customary to assume that all flow is confined, excepting those very special cases where the omission of a free surface would be an obvious violation, such as in earth dams and drawdown at wells.

To illustrate the effect that the prior assumption of confinement has on the solution, we shall investigate the flow around a single sheetpile embedded to a depth s in an infinitely deep layer of porous medium (Fig. 7-1b) and allow the tail-water boundary to develop without any prior restrictions.† This problem was solved previously in Sec. 5-2 (cf.

---

* Of course, any streamline may be taken as an impervious boundary [cf. Eq. (15), Sec. 4-4].

† The essentials of this solution were first given by Zhukovsky [164].

Fig. 5-9c), assuming the flow to be confined.   In the latter solution (Fig. 7-1a) all streamlines were required to intersect the tail-water boundary EF_∞ regardless of the magnitude of the head h_1.

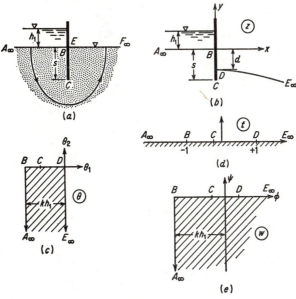

Fig. 7-1

The initial phases of the problem to be investigated were developed previously in Sec. 4-5 in connection with Zhukovsky's function.   The Zhukovsky θ plane was shown to be a semi-infinite strip (Fig. 7-1c) where $\theta = \theta_1 + i\theta_2 = w - ikz$.

Following through with the solution, we obtain from the Schwarz-Christoffel transformation for the mapping of the θ plane onto the t plane (Fig. 7-1d),

$$\theta = w - ikz = \frac{-kh_1}{\pi} \sin^{-1} t - \frac{kh_1}{2} \tag{1}$$

In like manner, the mapping of the w plane (Fig. 7-1e) onto the t plane gives

$$w = k(d + h_1) \sqrt{\frac{t+1}{2}} - kh_1 \tag{2}$$

which, after substitution into Eq. (1), yields

$$iz = (d + h_1) \sqrt{\frac{t+1}{2}} - \frac{h_1}{2} - \frac{h_1}{\pi} \sin^{-1} t \tag{3}$$

Now, noting at the bottom of the sheetpile (point C) that the velocity is infinite ($dw/dz = \infty$ or $dz/dw = 0$), upon differentiating Eqs. (1) and (2) with respect to t and dividing the first by the second, we get

$$1 - ik\frac{dz}{dw} = \frac{2\sqrt{2}}{\pi\sqrt{1-t}}\frac{h_1}{h_1 + d} \tag{4}$$

which at point C reduces to

$$t_C = 1 - \frac{8}{\pi^2}\left(\frac{h_1}{h_1 + d}\right)^2 \tag{5}$$

Also, at point C, $z = -is$; hence, substituting Eq. (5) into Eq. (3), we get

$$s = \sqrt{(h_1 + d)^2 - \frac{4h_1{}^2}{\pi^2}} - h_1 + \frac{2h_1}{\pi}\sin^{-1}\frac{2h_1/\pi}{h_1 + d} \tag{6}$$

A plot of Eq. (6) with d/h_1 as a function of s/h_1 is given in Fig. 7-2. It is seen from the figure that if the depth of embedment exceeds about

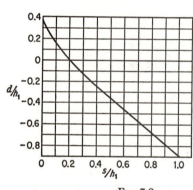

$$\frac{s}{h_1} = 1.1, \frac{d}{h_1} = 1.0$$

$$\frac{s}{h_1} = 2.07, \frac{d}{h_1} = 2.0$$

Fig. 7-2

22 per cent of the headwater height ($s/h_1 > 0.213$) the free surface will be entirely below EF in Fig. 7-1a. The exit gradient for this case, assuming confinement (Fig. 5-9c), is 1.49. Thus it appears that if the water ever reaches the tail-water boundary, piping will be imminent.

Finally, solving Eq. (1) for t and substituting the resulting expression into Eq. (2), we obtain

$$w = -k\left[h_1 - (h_1 \pm d)\cos\frac{\pi\theta}{2kh_1}\right] \tag{7}$$

which, noting that along the line of seepage $w = \phi = -ky$ and $\theta = -ikx$, yields for the equation of the free surface

$$y = h_1 - (h_1 \pm d)\cosh\frac{\pi x}{2h_1} \tag{8}$$

where the plus sign applies if d is below the x axis in Fig. 7-1b and the minus sign if d is above the axis.

Vedernikov [152] investigated the same problem, taking into account the effect of capillarity along the free surface. The resulting expressions are precisely the same as those given above, with $h_1 + h_c$ instead of h_1, where h_c is the height of capillary rise (Sec. 1-7). For example, for the equation of the free surface, Vedernikov gets

$$y = h_1 + h_c - (h_1 + h_c \pm d) \cosh \frac{\pi x}{2(h_1 + h_c)} \qquad (9)$$

7-3. Seepage through Homogeneous Earth Dam with Horizontal Underdrain

Recognizing the difficulties inherent in a closed-form solution to the general problem of flow through an earth dam, Nelson-Skornyakov [102] investigated two relatively simple cases for which he obtained exact solutions: (1) a dam with an horizontal upstream slope and (2) a dam with an upstream slope approximately vertical. Then, considering these solutions as limiting conditions, he presented a method whereby the solution could be approximated for any intermediate slope.

As was noted previously in Sec. 4-5, Nelson-Skornyakov used the modified form of Zhukovsky's function

$$G = z - \frac{iw}{k} = \left(x + \frac{\psi}{k}\right) + i\left(y - \frac{\phi}{k}\right) \qquad (1)$$

The advantage of this form of the function is primarily one of orientation. Whereas $y - \phi/k = 0$ along the free surface taking the vertical axis as positive down, with the form of Eq. (1), the image of this line will be along the real axis of the G plane.

Case 1. Horizontal Upstream Slope. The section to be considered in this case is shown in Fig. 7-3a. The corresponding w plane and G plane are shown in Fig. 7-3b and c, respectively.

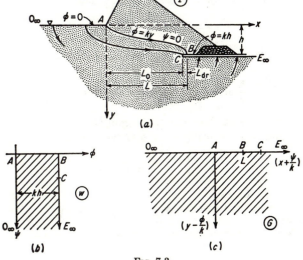

Fig. 7-3

From the Schwarz-Christoffel transformation we find that the mapping of the w plane onto the G plane is given by [cf. Eq. 383.2, Ref. 29]

$$G = L \sin^2 \frac{\pi w}{2kh} \tag{2}*$$

Substituting Eq. (2) into Eq. (1), we obtain the expression (called the *characteristic function*)

$$z = \frac{iw}{k} + L \sin^2 \frac{\pi w}{2kh} \tag{3}$$

which yields for the equations of any streamline and/or equipotential line

$$\begin{aligned}
x &= -\frac{\psi}{k} + \frac{L}{2}\left(1 - \cos\frac{\pi\phi}{kh}\cosh\frac{\pi\psi}{kh}\right) \\
y &= \frac{\phi}{k} + \frac{L}{2}\sin\frac{\pi\phi}{kh}\sinh\frac{\pi\psi}{kh}
\end{aligned} \tag{4}$$

Noting that along the free surface $w = ky$, $z = x + iy$, from Eq. (3) we find for the equation of the free surface

$$x = L \sin^2\left(\frac{\pi y}{2h}\right) \tag{5}$$

To determine the *length of underdrain*† L_{dr} (BC in Fig. 7-3a), we note that at point C the velocity is infinite ($dz/dw = 0$). Hence, differentiating Eq. (3) with respect to w,

$$\frac{dz}{dw} = \frac{i}{k} + \frac{\pi L}{2kh}\sin\frac{\pi w}{kh} \tag{6}$$

whence, at point C, where $w_c = kh + i\psi_c$, we find

$$w_C = kh + \frac{ikh}{\pi}\sinh^{-1}\frac{2h}{\pi L}$$

Substituting this expression for w, with $z = L_0 + ih$, into Eq. (3), we find that

$$L_{dr} = \frac{h}{\pi}\sinh^{-1}\frac{2h}{\pi L} + \frac{L}{2}\left[1 - \cosh\left(\sinh^{-1}\frac{2h}{\pi L}\right)\right] \tag{7}$$

For the exit gradient along the underdrain (BCE in Fig. 7-3a), inverting Eq. (6) and substituting $w = kh + i\psi$, we obtain

$$I_E = \left(1 - \frac{\pi L}{2h}\sinh\frac{\pi\psi}{kh}\right)^{-1} \tag{8}‡$$

---

* $\sin^2 A = \frac{1}{2}(1 - \cos 2A)$.

† This is a conventional definition and bears no relation to the actual physical underdrain length.

‡ See Prob. 4 for conservative method of estimating factor of safety with respect to piping.

where the corresponding value of x along the underdrain (where $\phi = kh$) is found from Eqs. (4) to be

$$x = -\frac{\psi}{k} + \frac{L}{2}\left(1 + \cosh\frac{\pi\psi}{kh}\right) \qquad (9)$$

Because of the implicit nature of Eqs. (8) and (9), it is expedient to assume a series of values for ψ, such as $\psi = 0.1\,kh$, $\psi = 0.2\,kh$, . . . , and then calculate the exit gradients from Eq. (8) and the corresponding values of x from Eq. (9).

Case 2. Upstream Slope Approximately Vertical. The section under consideration for this case is shown in Fig. 7-4a. Here the upstream boundary AO_∞ is taken as the line along which

$$\psi = -kx \qquad \phi = 0$$

where k is the coefficient of permeability. All other boundaries are the same as in case 1. Thus the only modification occurs in the G plane, where the image of the upstream surface will be $G_{AO} = iy$ (Fig. 7-4c). The w plane is shown in Fig. 7-4b.

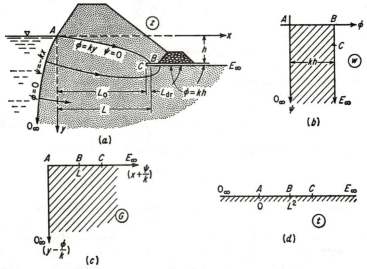

Fig. 7-4

We begin the solution by transforming the G plane onto the lower half of the t plane by $t = G^2$ (Fig. 7-4d). Then, mapping the w plane onto the t plane, it follows at once that the functional relationship between the z plane and the w plane is

$$z = \frac{iw}{k} + L\sin\frac{\pi w}{2kh} \qquad (10)$$

and the equation of the free surface is

$$x = L \sin \frac{\pi y}{2h} \tag{11}$$

Proceeding with the full development of the solution, the general expression for any streamline or equipotential line is

$$x = -\frac{\psi}{k} + L \sin \frac{\pi \phi}{2kh} \cosh \frac{\pi \psi}{2kh}$$

$$y = \frac{\phi}{k} + L \cos \frac{\pi \phi}{2kh} \sinh \frac{\pi \psi}{2kh} \tag{12}$$

the minimum length of underdrain is

$$L_{dr} = 2\left[\frac{L}{2} + \frac{h}{\pi} \sinh^{-1} \frac{2h}{\pi L} - \frac{L}{2} \cosh\left(\sinh^{-1} \frac{2h}{\pi L}\right)\right] \tag{13}$$

and the exit gradient along BCE is

$$I_E = \left(1 - \frac{\pi L}{2h} \sinh \frac{\pi \psi}{2kh}\right)^{-1} \tag{14}$$

where

$$x = -\frac{\psi}{k} + L \cosh \frac{\pi \psi}{2kh} \tag{15}$$

General Solution. We shall now consider the general form of the solution for any upstream slope intermediate to the two limiting cases considered above. On the basis of the results given in Eqs. (3) and (10), Nelson-Skornyakov assumed that the general expression for the functional relationship between the z plane and w plane could be written as

$$z = \frac{iw}{k} + L \sin^{2/n} \frac{\pi w}{2kh} \tag{16}$$

and the general expression for the free surface as

$$x = L \sin^{2/n} \frac{\pi y}{2h} \tag{17}$$

The parameter n in Eqs. (16) and (17) was shown to be 1 for the horizontal upstream slope and 2 for the case where the slope was almost vertical. For any intermediate slope, Nelson-Skornyakov assumed that n varies linearly between these two limiting values. For example, for an upstream slope at approximately 45° (1:1), $n = \frac{3}{2}$ and the characteristic function will be

$$z = \frac{iw}{k} + L \sin^{4/3} \frac{\pi w}{2kh} \tag{18a}$$

The equation of the free surface is then

$$x = L \sin^{4/3} \frac{\pi y}{2h} \tag{18b}$$

Following through with the solution for this case we would find that
the exit gradients (along BCE) would be given by

$$I_E = \left(1 - \frac{2\pi L}{3h} \sinh \frac{\pi\psi}{2hk} \cosh^{1/3} \frac{\pi\psi}{2kh}\right)^{-1} \tag{18c}$$

where $$x = -\frac{\psi}{k} + L \cosh^{2/3} \frac{\pi\psi}{2kh} \tag{18d}$$

To demonstrate the influence of various upstream surfaces on the free
surface, the curves for $n = 1$, $3/2$, and 2, with $L = 3h$, are shown in Fig.
7-5a. The shaded band on this figure denotes the usual range of slopes

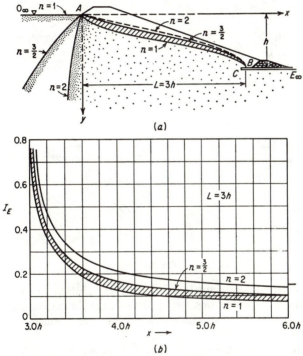

Fig. 7-5. (*After Nelson-Skornyakov* [102].)

for homogeneous dams without slope protection. Curves showing the
distribution of exit gradients along CE of the figure for the same three
n values are given in Fig. 7-5b.

7-4. Earth Structure with a Cutoff Wall*

The section to be investigated is shown in Fig. 7-6a. Cutoff walls are
generally used to lower the free surface and hence reduce the required

* Much of this section is based on a solution by Nelson-Skornyakov [102].

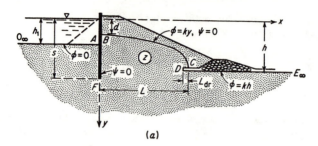

(a)

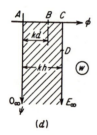

(b)

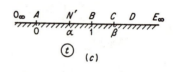

(c)

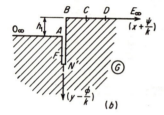

(d)

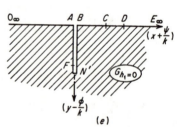

(e)

Fig. 7-6

width of the earth structure (dam). Also the solution to this problem can be taken as an approximation for the seepage characteristics through an homogeneous earth dam with an impervious core founded on a base of great depth.

The Zhukovsky G plane ($G = z - iw/k$) is shown in Fig. 7-6b. Applying the Schwarz-Christoffel transformation, the mapping of the G plane onto the t plane (Fig. 7-6c) is found to be [cf. Eq. 383.3, Ref. 29]

$$G = \frac{2h_1}{\pi(1 - 2\alpha)}\left[\sqrt{t^2 - t} + \frac{1}{2}(1 - 2\alpha)\cosh^{-1}(2t - 1)\right] \quad (1)$$

whence it follows immediately, from the definition of the G function, that the characteristic function is

$$z = \frac{iw}{k} + \frac{2h_1}{\pi(1 - 2\alpha)}\left[\sqrt{t^2 - t} + \frac{1}{2}(1 - 2\alpha)\cosh^{-1}(2t - 1)\right] \quad (2)$$

The function mapping the w plane (Fig. 7-6d) onto the t-plane is [cf. Eq. (2), Sec. 7-3]

$$t = \beta \sin^2 \frac{\pi w}{2kh} \tag{3}$$

where β is determined by the correspondence between the planes at points $B(t = 1, w = kd)$ as

$$\beta = \frac{1}{\sin^2 (\pi d/2h)} \tag{4}$$

A plot of Eq. (4) is given in Fig. 7-7a. Substituting Eq. (4) into Eq (3), we find that

$$t = \frac{\sin^2 (\pi w/2kh)}{\sin^2 (\pi d/2h)} \tag{5}$$

Considering the correspondence at points C ($t = \beta$, $z = L + ih$, and $w = \phi = kh$), it follows immediately from Eq. (2) that the parameter α

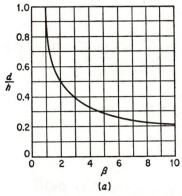

(a)

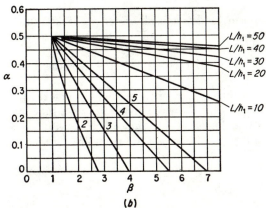

(b)

Fig. 7-7. (b) (*After Nelson-Skornyakov* [102].)

is given by the equation

$$\alpha = \frac{1}{2} - \frac{\sqrt{\beta^2 - \beta}}{\pi L/h_1 - \cosh^{-1}(2\beta - 1)} \tag{6}$$

A plot of α as a function of β and L/h_1 is given in Fig. 7-7b. Hence the functional relationship between the z plane and the w plane [Eq. (2)] is completely established.

To obtain the exit gradient along CDE (where $w = kh + i\psi$), we differentiate Eq. (2) with respect to w and substitute dt/dw from Eq. (3):

$$I_E = \left[1 - \frac{2\beta h_1(t - \alpha)}{h(1 - 2\alpha)} \frac{\sinh(\pi\psi/2kh)\cosh(\pi\psi/2kh)}{\sqrt{t^2 - t}} \right]^{-1} \tag{7a}$$

where the x distance along the boundary CDE is

$$x = -\frac{\psi}{k} + \frac{2h_1}{\pi(1 - 2\alpha)} \left[\sqrt{t^2 - t} + \frac{1}{2}(1 - 2\alpha)\cosh^{-1}(2t - 1) \right] \tag{7b}$$

and

$$t = \beta \cosh^2 \frac{\pi\psi}{2kh} \tag{7c}$$

As in Sec. 7-3, because of the implicit nature of Eqs. (7) it is advisable to assume a series of values for ψ, such as $\psi = 0.1\,kh, 0.2\,kh, \ldots$, and then determine the gradients from Eq. (7a) and the corresponding x values from Eq. (7b) with t as given in (7c).

Noting that the velocity at the bottom of the cutoff (point F) is infinite $(dz/dw = 0)$ and defining $w = \phi_F$ as the potential at that point with t given by Eq. (3), we find

$$\frac{2h_1}{h(1 - 2\alpha)} \frac{[\beta \sin^2(\pi\phi_F/2kh) - \alpha]\cos(\pi\phi_F/2kh)}{\sqrt{\dfrac{1}{\beta} - \sin^2(\pi\phi_F/2kh)}} = 1 \tag{8a}$$

from which ϕ_F can be determined. Once ϕ_F is known, the depth of embedment of the cutoff wall (s in Fig. 7-6a) can be found from Eq. (2).

$$\frac{s}{h} = \frac{\phi_F}{kh} + \frac{2h_1}{h(1 - 2\alpha)\pi} \left[\sqrt{t^2 - t} + \frac{1}{2}(1 - 2\alpha)\cosh^{-1}(2t - 1) \right] \tag{8b}$$

where only the imaginary part of the second term on the L.H.S. applies, and $t = \beta \sin^2(\pi\phi_F/2kh)$. Curves giving s/h as a function of d/h and L/h for $h_1/h = 0$ and $h_1/h = 1$ are shown in Fig. 7-8.

Finally, from Eq. (2) with $w = \phi = ky$, we find for the equation of the free surface

$$x = \frac{2h_1}{\pi(1 - 2\alpha)} \left[\sqrt{t^2 - t} + \frac{1}{2}(1 - 2\alpha)\cosh^{-1}(2t - 1) \right] \tag{9}$$

where

$$t = \beta \sin^2 \frac{\pi y}{2h} \geqq 1 \qquad d \leqq y \leqq h$$

From Fig. 7-8 we note that for $L/h > 2$ and $s/h > 1.5$ the influence of the elevation of the upstream boundary h_1 on the *height of rise d* (and hence the free surface) is small. Therefore, for these conditions a great simplification can be effected in the analysis by assuming $h_1 = 0$.

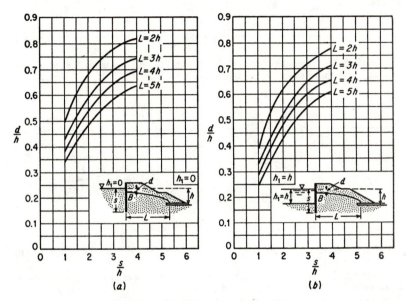

FIG. 7-8. (*After Nelson-Skornyakov* [102].)

Approximate Solution, $h_1 = 0$. Taking $h_1 = 0$ in the equations above,* we obtain ($\gamma = \pi d/2h$):

1. The characteristic function,

$$z = \frac{iw}{k} + \frac{L \sin (\pi w/2kh)}{\cos \gamma} \sqrt{\sin^2 \frac{\pi w}{2kh} - \sin^2 \gamma} \qquad (10)$$

2. The equation of the free surface,

$$x = \frac{L \sin (\pi y/2h)}{\cos \gamma} \sqrt{\sin^2 \frac{\pi y}{2h} - \sin^2 \gamma} \qquad (11)$$

3. The depth of embedment of the cutoff wall,

$$\frac{s}{h} = \frac{L \sin^2 \gamma}{2h \cos \gamma} + \frac{2}{\pi} \sin^{-1} \left(\frac{\sin^2 \gamma}{2 \cos \gamma \sqrt{1 + \sin^4 \gamma/4 \cos^2 \gamma} + \sin^2 \gamma} \right)^{\frac{1}{2}} \qquad (12)$$

* Note that α is a function of h_1 and that Eq. (6) for $h_1 = 0$ yields $2h_1/(1 - 2\alpha) = L\pi/\sqrt{\beta^2 - \beta}$.

4. The exit gradient along CDE,

$$I_E = \left[1 - \frac{\pi L(2t - 1) \sinh (\pi\psi/kh)}{4h \cos \gamma \sqrt{t^2 - t}} \right]^{-1} \tag{13a}$$

where

$$t = \frac{\cosh^2 (\pi\psi/2kh)}{\sin^2 \gamma}$$

$$x = \frac{L \cosh (\pi\psi/2kh)}{\cos \gamma} \sqrt{\cosh^2 \frac{\pi\psi}{2kh} - \sin^2 \gamma} - \frac{\psi}{k} \tag{13b}$$

Example 7-1. Given the section shown in Fig. 7-9 with $s = 60$ ft, $h = 40$ ft, and $L = 2h$. Determine the pertinent flow characteristics.

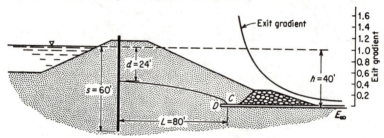

FIG. 7-9. (*After Nelson-Skornyakov* [102].)

For $s/h = 60/40$, with $L/h = 2$, from Fig. 7-8a ($h_1 = 0$), we obtain $d/h = 0.6$; hence $d = 24$ ft.

The free surface is obtained from Eq. (11) by assuming a sequence of values of y/h. Some values are given in Table 7-1. The resulting free surface is shown in Fig. 7-9. Some values for the exit gradient along the underdrain are given in Table 7-1. The distribution of the exit gradients is shown in Fig. 7-9.

Table 7-1*

Free surface

$\dfrac{y}{h}$	0.6	0.8	0.9	1.0
$\dfrac{x}{h}$	0	1.6	1.9	2.0

Exit gradient

ψ	0.2kh	0.5kh	kh
I_E	−1.6	−0.09	−0.02
$\dfrac{x}{h}$	2.2	4.1	19.0

* From F. B. Nelson-Skornyakov, "Seepage in Homogeneous Media," Gosudarctvennoe Izd. Sovetskaya Nauka, Moscow, 1949.

A set of equations similar to Eqs. (10) to (13) can be obtained for $h_1 = h$ (see Prob. 6). The solutions for these two cases ($h_1 = 0$ and $h_1 = h$) may then be taken as the limiting conditions for any intermediate upstream slope.

For the minimum length of underdrain L_{dr}, Nelson-Skornyakov [102] demonstrated that it is reasonable to assume for $L > 3h$ that $L_{dr} < 0.15\,h$ (this will be demonstrated in Sec. 7-6).

7-5. Earth Structure with a Cutoff Wall at the Toe

Noting the similarity between the z plane for the section shown in Fig. 7-10a and Fig. 7-6a with $h_1 = 0$ (one being the reverse of the other),

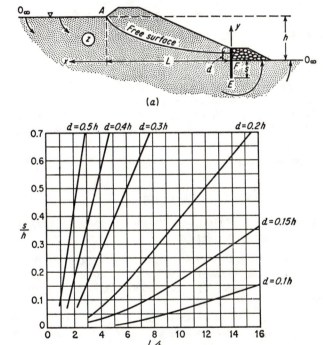

Fig. 7-10. (*After Nelson-Skornyakov* [102].)

Nelson-Skornyakov gave for the former the characteristic function

$$z = -\frac{iw}{k} + \frac{iL \sin (\pi w/2kh)}{\cos \gamma} \sqrt{\sin^2 \gamma - \sin^2 \frac{\pi w}{2kh}} \qquad (1)$$

where $\gamma = \pi d/2h$. Hence we have immediately for the equation of the free surface

$$x = \frac{L \sin (\pi y/2h)}{\cos \gamma} \sqrt{\sin^2 \frac{\pi y}{2h} - \sin^2 \gamma} \qquad (2)$$

For the depth of embedment of the cutoff, noting that the velocity at point E is unbounded, and proceeding as before, we find that

$$\frac{s}{h} = \frac{L}{h} \sin \left(\frac{\pi \phi_E}{2kh} \right) \sqrt{\alpha} - \frac{\phi_E}{kh} \qquad (3a)$$

where the parameter α is given by the expression

$$\frac{L}{h} = \frac{\sqrt{\alpha}}{\pi \cos \gamma (\alpha - \frac{1}{2} \tan^2 \gamma) \sqrt{1 + \alpha}} \qquad (3b)$$

and ϕ_E is given by

$$\cos \frac{\pi \phi_E}{2kh} = \sqrt{1 + \alpha} \cos \gamma \qquad (3c)$$

Nelson-Skornyakov also gave the nomograph of Fig. 7-10b, from which the relationships between the various dimensional parameters of the section are easily obtained.   We note from this plot that for ratios of $L/h >$ 10 great depths of embedment are required to raise the downstream water level d significantly.

The exit gradient at point F, the maximum gradient along FO, is

$$I_E = - \left(1 - \frac{\pi L}{2h} \tan \gamma \right)^{-1} \qquad (4)$$

7-6. Earth Structure with Horizontal Drain Underlain by Impervious Material of Infinite Extent

Again in considering this problem, Nelson-Skornyakov investigates the two limiting cases: (1) a structure with an upstream slope close to the vertical (Fig. 7-11a) and (2) a dam with an horizontal upstream slope (Fig. 7-13).

Case 1.   Close to Vertical Upstream Slope.   With the upstream slope as an equipotential line along which $\psi = -kx$ and the streamline along the vertical impervious boundary as $\psi = q$, the corresponding G plane and w plane are as shown in Figs. 7-11b and c, respectively.   We note that, although the structure is of infinite extent, as the flow is confined between the free surface ($\psi = 0$) and the streamline $\psi = q$, the discharge quantity will be finite.

The mapping of the G plane onto the auxiliary p plane (Fig. 7-11d) is given by [cf. Eq. (2), Sec. 7-3]

$$p = \sin^2 \frac{\pi G}{2(L_0 + q/k)} \qquad (1)$$

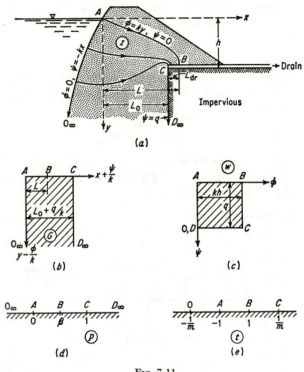

FIG. 7-11

and of the w plane onto the t plane [cf. Eq. (5), Sec. 5-4] by

$$t = \mathrm{sn}\left(\frac{2wK}{kh} - K\right) \qquad (2)$$

Now mapping the t plane onto the p plane by the cross-ratio formula [Eq. (4), Sec. 3-7] for the triples at O, A, and C, we see that

$$p = \frac{2m(1 + t)}{(1 + m)(1 + mt)} \qquad (3)$$

Substituting this expression for p into Eq. (1), from the definition of the G plane ($G = z - iw/k$), we have for this case the characteristic function

$$z = \frac{iw}{k} + \frac{2}{\pi}\left(L_0 + \frac{q}{k}\right)\sin^{-1}\sqrt{\frac{2m(1 + t)}{(1 + m)(1 + mt)}} \qquad (4)$$

where t and w are related by Eq. (2).

From the correspondence between the t and w planes at point C, it follows at once that the discharge is

$$q = \frac{khK'}{2K} \tag{5}$$

Whereas the velocity at point C is infinite as in previous sections, differentiating Eq. (4) with respect to w and setting the resulting expression equal to zero, we find that

$$\frac{L_0}{h} = \frac{\pi}{2K(1 - m)} - \frac{K'}{2K} \tag{6}$$

whence the modulus m can be determined for any L_0/h, and consequently the discharge can be obtained from Eq. (5). A plot of the discharge as a function of L_0/h is given in Fig. 7-12a.

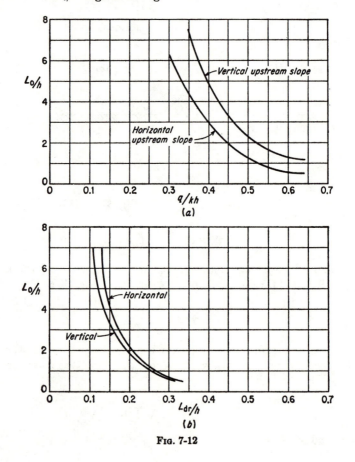

Fig. 7-12

From the correspondence at points B ($z = L + ih$, $t = 1$, $w = kh$) and Eqs. (5) and (6), we obtain without difficulty from Eq. (4)

$$\frac{L}{h} = \frac{1}{K(1 - m)} \sin^{-1} \frac{2 \sqrt{m}}{1 + m} \tag{7}$$

and hence the minimum length of underdrain, $L_{dr} = L - L_0$, is

$$\frac{L_{dr}}{h} = \frac{K'}{2K} - \frac{1}{K(1 - m)} \cos^{-1} \frac{2 \sqrt{m}}{1 + m} \tag{8}$$

A plot of Eq. (8) with L_{dr}/h as a function of L_0/h is given in Fig. 7-12b.*

The equation of the free surface is readily shown from Eqs. (4) and (6) to be

$$x = \frac{h}{K(1 - m)} \sin^{-1} \sqrt{\frac{2m(1 + \theta)}{(1 + m)(1 + m\theta)}} \tag{9}$$

where $\theta = \operatorname{sn} (2yK/h - K)$.

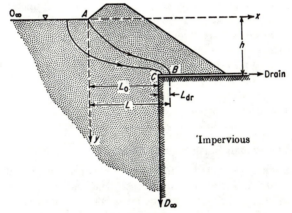

FIG. 7-13

Case 2. Horizontal Upstream Slope. Proceeding as above for the limiting case of a horizontal upstream slope (Fig. 7-13), Nelson-Skornyakov found:

1. The characteristic function,

$$z = \frac{iw}{k} - \left(L_0 + \frac{q}{k}\right) \left[\sqrt{\frac{(1 - m)(1 - mt)}{(1 + m)(1 + mt)}} - 1 \right] \tag{10}$$

where $t = \operatorname{sn} (2wK/kh - K)$.

2. The discharge,

$$q = \frac{khK'}{2K} \tag{11}$$

---

* The plot given by Nelson-Skornyakov for Eq. (8) was found to be in error.

where the modulus m is given by

$$\frac{L_0}{h} = \frac{1}{K(1 - m)} - \frac{K'}{2K} \tag{12}$$

3. The parameter L,

$$\frac{L}{h} = \frac{2m}{K(1 - m^2)} \tag{13}$$

4. The minimum length of underdrain,

$$\frac{L_{dr}}{h} = \frac{K'}{2K} - \frac{1}{K(1 + m)} \tag{14}$$

Plots of the discharge and the minimum length of underdrain for the horizontal case are shown in Fig. 7-12.*

Figure 7-12b shows that for $L_0/h > 3$, the minimum length of underdrain L_{dr} is less than $0.2h$ for any shape or inclination of upstream slope. Whereas the infinite extent of the impervious base underlying the underdrain, such as given here, would require a greater *minimum length of underdrain* than the same structure without the impervious boundary, the lengths given in Fig. 7-12b can be considered to be on the safe side for any homogeneous section of great depth with any shape or inclination of upstream slope. Also, noting that the impervious boundary may be taken as a cutoff wall or sheetpile, the above results can be considered as a reasonable approximation for these cases when the cutoff is complete.

7-7. Seepage through Earth Structures into Drains of Finite Length

In the previous sections of this chapter it was assumed that the flow terminated into drains of infinite extent. We shall now consider these drains to be finite. The section to be investigated is shown in Fig. 7-14a. Point D on this figure locates the terminal point of what may be called the *critical* streamline (or *water divide*). All streamlines included between the free surface AB and the critical streamline terminate at the drain; those below terminate at the ground surface to the right of point D. The location of this point and the locus of the free surface DB' are of considerable engineering importance in that the effects of seepage must be considered in the design of the right cut slope as well as the left.

The essential analytical results of this section were first derived in a somewhat modified way by Nelson-Skornyakov; however, his results are not amenable to design use in that they contained elliptic integrals of the third kind in implicit form. To obviate this difficulty Harr and Brahma [52] reduced these solutions to a simple graphical form given below.

* See previous footnote.

As in the previous sections, the two limiting cases will be investigated: (1) horizontal upstream slope and (2) close to vertical upstream slope.

Case 1. Horizontal Upstream Slope (Fig. 7-14b). The G plane is shown in Fig. 7-14c and the corresponding w plane in Fig. 7-14d. Select-

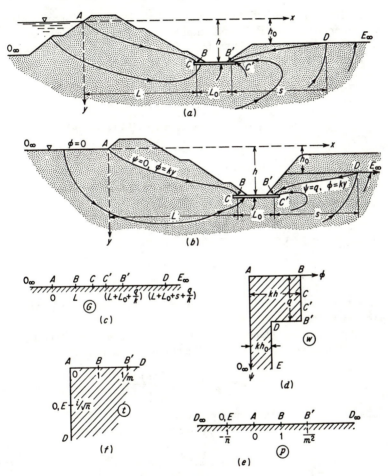

Fig. 7-14

ing a parametric p plane as shown in Fig. 7-14e, we have for the mapping of the w plane onto the p plane ($N = 0$, $w_A = p_A = 0$)

$$w = M \int_0^p \frac{dp}{(p + 1/n) \sqrt{p(p - 1)(p - 1/m^2)}}$$

Now, introducing a change of variable $t^2 = p$ (this is equivalent to defining the t plane as given in Fig. 7-14f), we obtain

$$w = M \int_0^t \frac{2t\,dt}{(t^2 + 1/n)\sqrt{t^2(t^2 - 1)(t^2 - 1/m^2)}} = M_1\Pi(m,n,t) \quad (1)$$

where $\Pi(m,n,t)$ is the elliptic integral of the third kind with modulus m, parameter n (cf. Sec. B-6), and amplitude t. From the correspondence at points B ($w = kh$, $t = 1$) we find $M_1 = kh/\Pi_0(m,n)$, where $\Pi_0(m,n)$ is the complete elliptic integral of the third kind, and

$$w = kh \frac{\Pi(m,n,t)}{\Pi_0(m,n)} \quad (2)$$

At points B', $w = kh + iq$, $t = 1/m$, and Eq. (2) becomes*

$$kh + iq = \frac{kh}{\Pi_0(m,n)}\left[\Pi_0(m,n) + \frac{im^2}{n + m^2}\Pi_0(m',n')\right]$$

where
$$n' = -\frac{nm'^2}{m^2 + n}$$
$$m' = \sqrt{1 - m^2}$$

and hence we get for the quantity of seepage,

$$q = \frac{khm^2\Pi_0(m',n')}{(n + m^2)\Pi_0(m,n)} \quad (3)$$

Thus the discharge can be determined once the modulus m and the parameter n are known.

Mapping the G plane onto the p plane (cross-ratio formula) and substituting $t^2 = p$, we obtain for the characteristic function

$$z = \frac{iw}{k} + L\frac{(1 + n)t^2}{(1 + nt^2)} \quad (4)$$

Substituting the corresponding values at points B' into Eq. (4) ($z = L + L_0 + ih$, $w = kh + iq$, and $t = 1/m$), we have

$$L_0 = L\frac{m'^2}{m^2 + n} - \frac{q}{k} \quad (5)$$

In a similar manner, at points D, where $z = L + L_0 + s + ih_0$, $w = kh_0 + iq$, and $t = \infty$, we obtain

$$L_0 = \frac{L}{n} - \frac{q}{k} - s \quad (6)$$

* See Eqs. (28), Appendix B.

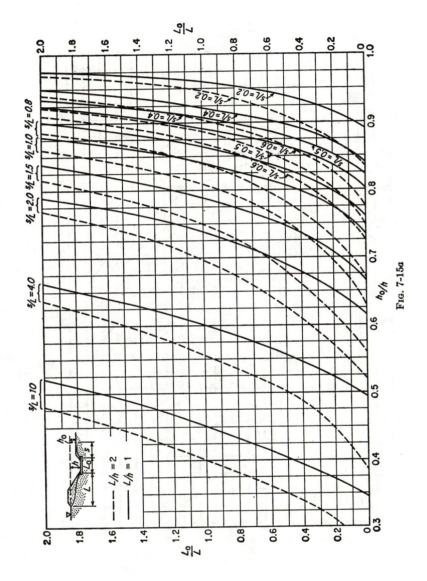

Fig. 7-15a

189

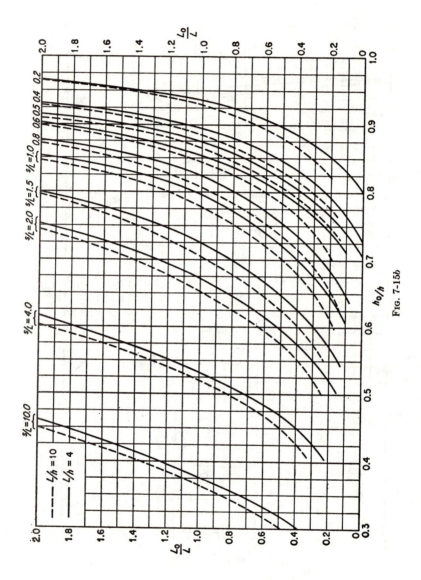

Fɪɢ. 7-15b

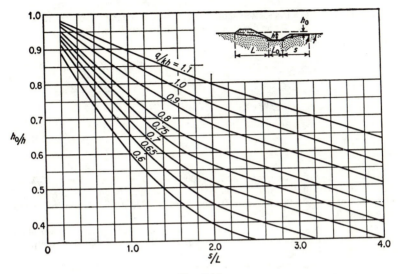

Fig. 7-15c

which, when combined with Eq. (5), yields

$$\frac{s}{L} = \frac{n+1}{n + n^2/m^2} \tag{7}$$

Now, substituting the corresponding values of the w and t planes at points D into Eq. (2),

$$kh_0 + iq = \frac{kh}{\Pi_0(m,n)} \int_0^\infty \frac{dt}{(1 + nt^2)\sqrt{(1 - t^2)(1 - m^2t^2)}}$$

Using the results given in Eqs. (27b) and (28), Appendix B, we obtain

$$\frac{h_0}{h} = \frac{\pi}{2\Pi_0(m,n)\sqrt{(1 + n)(1 + m^2/n)}} \tag{8}$$

Equations (3), (5), (7), and (8) yield the complete solution of the problem, assuming the characteristic dimensions L_0/h, L, and h_0/h of the flow domain are known. With regard to L_0: as a first approximation L_0 can be taken as the total width of the drain. This will generally suffice; however, should greater precision be required, L_0 could be taken as the width of the drain reduced by twice $0.1\,h$ with L increased by $0.1\,h$. The curves given in Fig. 7-15 present the above results in simple graphical form. The use of these curves will be illustrated by an example.

Example 7-2. For the section shown in Fig. 7-16, assuming the upstream slope as horizontal determine (a) the distance s and (b) the reduced discharge.

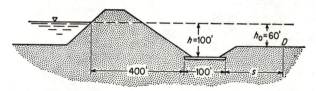

Fig. 7-16

As a first approximation, we take $L = 400$ ft and $L_0 = 100$ ft. Hence, $L/h = 4$, $L_0/L = 0.25$, and $h_0/h = 0.6$. From Fig. 7-15b we find (solid lines for $L/h = 4$) $s/L = 1.4$ or $s = 560$ ft, and point D (Fig. 7-16) is 1,060 ft from the entrance point ($x = 1,060$ ft).

For the discharge, from Fig. 7-15c, with $s/L = 1.4$ and $h_0/h = 0.6$, we obtain $q/kh = 0.70$; hence the reduced discharge is $q/k = 70$ ft (m and n in this example are 0.7 and 0.5, respectively).

Taking into account the difference between the width of the drain and L_0, assuming $L_{dr} = 0.1\ h$, $L_0 = 80$ ft, and $L = 410$, we obtain $s = 520$ ft, and point D is 1,010 ft from the entrance point. The reduced discharge for this case would be $q/k = 66$ ft.

Case 2. Close to Vertical Upstream Slope (Fig. 7-17a). With the upstream slope as the equipotential line ($\phi = 0$) along which $\psi = -kx$,

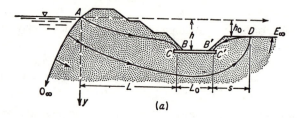

(a)

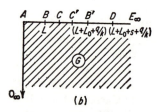

(b)

Fig. 7-17

the corresponding G plane is as given in Fig. 7-17b. The w plane for this case remains the same as in case 1 above (Fig. 7-14d). With the parametric t plane as before (Fig. 7-14f), the mapping of the w plane onto the t plane will be given by Eq. (2). Hence Eq. (3) will provide the discharge quantity, and Eq. (8) will be valid for the ratio of h_0/h for the present case as well as for the case of the horizontal upstream slope.

The appropriate mapping of the G plane onto the t plane is

$$G = L \sqrt{\frac{(1+n)t^2}{1+nt^2}} \tag{9}$$

and hence the characteristic function will be given by

$$z = \frac{iw}{k} + L \sqrt{\frac{(1+n)t^2}{1+nt^2}} \tag{10}$$

Substituting the corresponding values of the G plane and t plane at points B' ($G = L + L_0 + q/k$ and $t = 1/m$) into Eq. (10), we obtain

$$L_0 = L\left(\sqrt{\frac{1+n}{m^2+n}} - 1\right) - \frac{q}{k} \tag{11}$$

Now, making use of the conditions at points D, where

$$G = L + L_0 + s + \frac{q}{k}$$

and $t = \infty$, from Eqs. (9) and (11) we find that for the close to vertical upstream slope

$$\frac{s}{L} = \sqrt{\frac{1+n}{n}} - \sqrt{\frac{1+n}{m^2+n}} \tag{12}$$

As in the previous case Eqs. (3), (8), (11), and (12) provide the complete solution for the close to vertical upstream slope. The results are reduced to graphical form in Fig. 7-18.

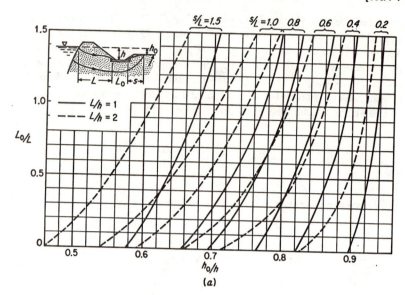

(a)

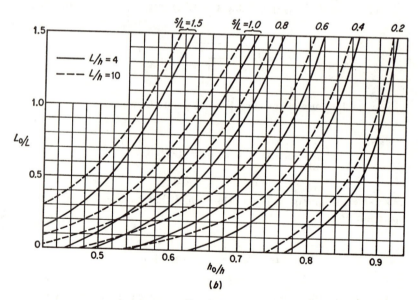

(b)

Fig. 7-18

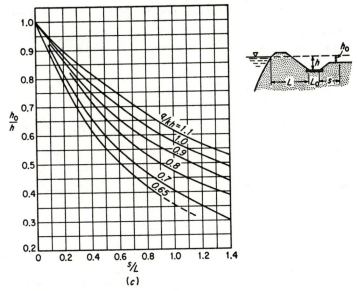

Fɪɢ. 7-18 (*continued*)

Finally, it should be noted that although the loci of the free surfaces can be obtained from Eqs. (4) and (10), a great simplification can be effected by taking these surfaces as parabolas between the points A-B and D-B'.

PROBLEMS

1. An observation well is to be located as shown in Fig. 7-19. Estimate the elevation of the free surface at the well.

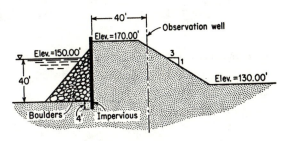

Fɪɢ. 7-19

2. Solve Prob. 1 if the porous medium is fine sand with a height of capillary rise of 2 ft.

3. Obtain the equivalent forms of Eqs. (18), Sec. 7-3, for a dam with an upstream slope of 30°.

4. The dam shown in Fig. 7-20 rests on a very deep bed of homogeneous soil with $k = 0.03$ cm/sec. Determine the pertinent dimensions of the dam and the locus of

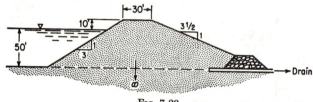

Fig. 7-20

the free surface assuming that the factor of safety with respect to piping along the underdrain can be taken as [see Eq. (2), Sec. 5-6]

$$\text{F.S.} = \frac{I_{cr}}{I_E} + \frac{2H\gamma_e}{h\gamma_w}$$

where H is the height of the overlying soil (filter and/or embankment) with a unit weight $\gamma_e = 2\gamma_w$, and h is the total head loss through the dam.

5. Derive Eqs. (10) to (13), Sec. 7-4, taking the G plane for $h_1 = 0$ as shown in Fig. 7-6e.

6. Derive the form of Eqs. (10) to (13), Sec. 7-4, for $h_1 = h$.

7. Obtain the locus of the free surface for the section shown in Fig. 7-21 if $x = 50$ ft. If the ground surface (CD_∞) is 6 ft below the upstream water elevation, estimate the maximum x distance that would prevent tail water from rising above point C.

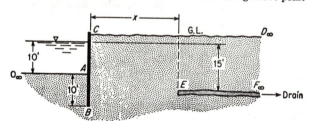

Fig. 7-21

8. Discuss design considerations for the section shown in Fig. 7-22.

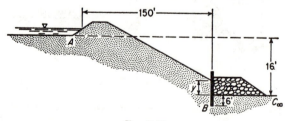

Fig. 7-22

9. Verify Eqs. (10) to (14), Sec. 7-6.

10. Estimate the reduced seepage and determine the approximate location of the free surface for the section shown in Fig. 7-23.

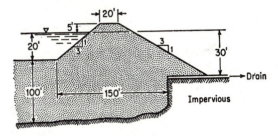

Fig. 7-23

11. Discuss the significance of the solution in Sec. 7-7 if (a) $h_0 = h$; (b) $L_0 \approx 0$.

12. Estimate the reduced seepage for the section shown in Fig. 7-24 and obtain the locus of the free surfaces.   The water divide is at point B.

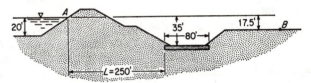

Fig. 7-24

13. Solve Prob. 12 if the water divide (point B) is observed 700 ft from point A and the length L is not known.

8

Unconfined Flow through Homogeneous Earth Structures of Finite Depth

8-1. Introduction

In the previous chapter various unconfined flow systems were investigated and the characteristic relationships were obtained assuming an infinite depth of porous media. If we recall that any streamline can be taken as an impervious boundary, we may extend the same solutions to include flow systems with finite lower impervious boundaries where one of the streamlines coincides (approximately) with an impermeable foundation. Some aspects of unconfined flow through shallow foundations were examined previously in Chap. 2 using Dupuit's assumption and in Chap. 3 by generalizing Kozeny's solution for a parabolic upstream slope. We shall consider in this chapter the application of other analytical methods to the solution of problems of this type.

8-2. Solution by Inversion

The fundamentals of inversion were developed previously in Sec. 3-6. It was shown that a transformation of the form $w = 1/z$ takes circles passing through the origin in the z plane into straight lines in the w plane. Also, straight lines passing through the origin in the z plane remain straight lines in the w plane. The inversion process coupled with the velocity hodograph (Sec. 4-1) provides the means of solving a variety of unconfined flow problems. To illustrate the method, we shall consider again the seepage through an homogeneous dam with a close to vertical upstream slope (Fig. 7-4a). The velocity hodograph* of this section will be taken as shown in Fig. 8-1a; that is, all boundaries of the hodograph are known with the exception of the upstream surface OA which, for simplicity, is taken as a circular arc of diameter k (coefficient of permeability). The conjugate of the hodograph plane, the dw/dz plane,

* Note that the y axis was taken positive vertically down; hence the vertical component of the velocity, as shown in Fig. 8-1a and b, is also positive vertically down.

is shown in Fig. 8-1b. By inversion, the dz/dw plane is found to be as shown in Fig. 8-1c. For example, the equation of the semicircle OAB in the hodograph plane, $u^2 + v^2 + kv = 0$, has as its inverse [Eq. (2), Sec. 3-6] the straight line wherein the imaginary part of $dz/dw = 1/k$.

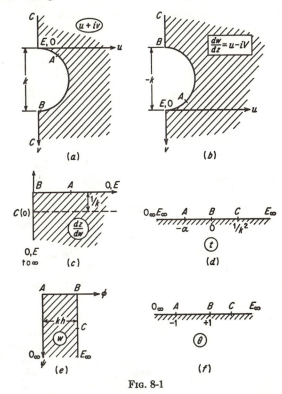

Fig. 8-1

Selecting a parametric t plane as shown in Fig. 8-1d, we have immediately for the mapping of the dz/dw plane onto the t plane

$$t = -\left(\frac{dz}{dw} - \frac{i}{k}\right)^2 \tag{1}$$

The w plane is reproduced in Fig. 8-1e. For the mapping of the w plane onto the θ plane of Fig. 8-1f we find

$$\theta = -\cos\frac{\pi w}{kh} \tag{2}$$

Now, noting that the θ plane and t plane are linearly related by (M is a constant)

$$t = M(\theta - 1) \tag{3}$$

and substituting for t and θ into Eq. (1), we obtain

$$\frac{dz}{dw} = \frac{i}{k} + M_1 \cos \frac{\pi w}{2kh} \qquad (4)*$$

Integrating Eq. (4) and considering the correspondence between the planes at points B ($z = L + ih$, $w = kh$), we find $M_1 = \pi L/2kh$ and

$$z = \frac{iw}{k} + L \sin \frac{\pi w}{2kh} \qquad (5)$$

which is precisely the expression we obtained previously in Sec. 7-3 [Eq. (10)].

8-3. Rockfill Dams with Central Cores without Tail Water

In most rockfill dams, such as the Cherry Valley Dam [1] shown in Fig. 8-2a, the head loss through the rock slopes is negligible in comparison to the loss through the relatively impermeable core. For such dams we need investigate only the seepage characteristics of the cores. We shall begin our discussion by considering a particular case, of some interest, for which a closed-form solution will be obtained. The section to be investigated is a symmetrical triangular core with base angles of 45° and a full head of water, as shown in Fig. 8-2b. The assumption of 1:1 slopes and of a full head, although conservative, is not extravagant,

Table 8-1. Some Dimensions on Three Rockfill Dams*

Dam	Core slopes‡		Elevation of core top, ft	Maximum design head-water, ft	Normal maximum head-water, ft
	Upstream	Down-stream			
Nottely (184)†.........	1:1	1.2:1	1,794	1,788	1,780
Watauga (318).........	0.85:1	0.85:1	1,998	1,988	1,959
South Holston (285)....	0.75:1	0.75:1	1,765	1,755	1,729

* From "Rockfill Dams: Performance of TVA Central Core Dams," by G. K. Leonards and O. H. Raines; ASCE "Symposium on Rockfill Dams," Proc. Symposium Series no. 3, reprinted from *J. Power Div.*, September, 1958.

† Figure in parenthesis is core height above impervious base.

‡ First figure is horizontal distance.

as shown by the data in Table 8-1 for three rockfill dams with central cores built by the Tennessee Valley Authority (cf. paper 1,736 of Ref. 1).

* Note that $[1 + \cos (\pi w/kh)]/2 = \cos^2 (\pi w/2kh)$.

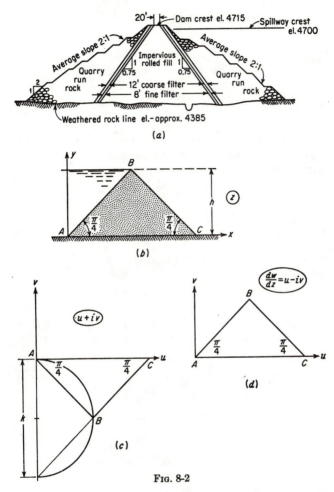

FIG. 8-2

Noting that the z plane in Fig. 8-2b and the dw/dz plane in Fig. 8-2d are similar triangles, we write

$$\frac{dw}{dz} = Cz \tag{1}*$$

where C is a constant of proportionality. Now, considering the correspondence at points B, where $z = h + ih$ and $dw/dz = (k + ik)/2$, we find $C = k/2h$ and

$$\frac{dw}{dz} = \frac{kz}{2h} \tag{2}$$

* This was first given by Davison [26] in 1937.

Integrating this expression we get

$$w = \frac{kz^2}{4h} - kh$$

which, after separation into real and imaginary parts, yields for the equations of the equipotential lines and streamlines

$$\phi = \frac{k}{4h}(x^2 - y^2) - kh \qquad \psi = \frac{kxy}{2h} \tag{3}$$

Whereas the bottom streamline ($y = 0$) is $\psi = 0$, the streamline through $x = y = h$ will be q, and hence from the second of Eqs. (3) we obtain for the discharge through the section:

$$q = \frac{kh}{2} \tag{4}$$

For the pressure in the water along the base of the dam AC, from the first of Eqs. (3) with $y = 0$, we find

$$\phi = \frac{kx^2}{4h} - kh$$

whence
$$p_{AC} = \gamma_w \left(h - \frac{x^2}{4h} \right) \tag{5}$$

The result given in Eq. (4) can also be established directly from Darcy's law and the assumption of horizontal flow [see Eq. (14), Sec. 2-7]. To

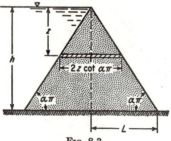

FIG. 8-3

illustrate, writing $q = kIA$, from Fig. 8-3 we see that $A = h$ and $I = (2 \cot \alpha\pi)^{-1}$; hence

$$q = \frac{kh}{2} \tan \alpha\pi \tag{6}$$

which for $\alpha = \frac{1}{4}$ yields identically Eq. (4).

Another section for which the exact value of the quantity of seepage* can be obtained is that of core with a vertical upstream slope with or

---

* This solution was first derived by Mkhitarian [93, 94]. In his solution Eq. (11) was given as the sum of an infinite series which, in the present development, is reduced to a tabulated function.

without tail water (Fig. 8-4a).   Referring to Fig. 8-4a, we specify: along
EF, $\phi = -kh$; along GK, $\phi = -kh_0$; for the surface of seepage EG, $\psi = q$
and $\phi = -ky$; and along FK, $\psi = 0$.   The resulting w plane is shown in
Fig. 8-4d, the hodograph plane in Fig. 8-4b.   For this case we see that

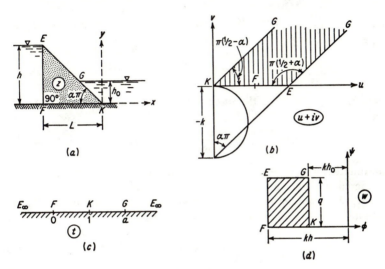

FIG. 8-4

the region of flow in the hodograph plane lies entirely in the first quadrant
(shown shaded).

Selecting an auxiliary t plane as shown in Fig. 8-4c, we have for the
mapping of the z plane onto this plane

$$z = M \int_1^t (t - 1)^{\alpha-1} t^{-\frac{1}{2}}\, dt \tag{7}$$

From the correspondence between planes at points E ($x = -h \cot \alpha\pi$,
$y = ih$, $t = \infty$), we get

$$M = \frac{h(i - \cot \alpha\pi)}{J} \tag{8}$$

where $$J = \int_1^\infty (t - 1)^{\alpha-1} t^{-\frac{1}{2}}\, dt$$

Substituting $t = 1/p$ into the expression for J, we see that J is the beta
function $B(\frac{1}{2} - \alpha, \alpha)$ [cf. Eq. (30), Appendix B], or in terms of gamma
functions

$$J = \frac{\Gamma(\frac{1}{2} - \alpha)\Gamma(\alpha)}{\Gamma(\frac{1}{2})} \tag{9}$$

Hence J and M can be determined for any α. Considering the correspondence at points G [$z = h_0(i - \cot \alpha\pi)$, $t = a$], we find

$$\frac{h_0}{h} = \frac{J'}{J} \tag{10}$$

where $J' = \int_1^a (t - 1)^{\alpha-1} t^{-\frac{1}{2}} dt$. Substituting $t = 1/p$ into the expression for J', we find

$$J' = J - \int_0^{1/a} (1 - p)^{\alpha-1} p^{-\alpha-\frac{1}{2}} dp = J - \mathrm{B}_{1/a}(\tfrac{1}{2} - \alpha, \alpha)$$

where $\mathrm{B}_{1/a}(\tfrac{1}{2} - \alpha, \alpha)$ is the *incomplete beta function*.* Hence Eq. (10) becomes

$$\frac{h_0}{h} = 1 - \frac{\mathrm{B}_{1/a}(\tfrac{1}{2} - \alpha, \alpha)}{\mathrm{B}(\tfrac{1}{2} - \alpha, \alpha)} = 1 - I_{1/a}(\tfrac{1}{2} - \alpha, \alpha) \tag{11}$$

The function $I_{1/a}(\tfrac{1}{2} - \alpha, \alpha)$ is well tabulated;* consequently, a can be considered as a known quantity for any h_0/h and α.

Now, the mapping of the dw/dz plane (the conjugate of the shaded portion of Fig. 8-4b) onto the t plane is given by

$$\frac{dw}{dz} = A \int_1^t (t - 1)^{-\frac{1}{2}-\alpha}(t - a)^{-1} dt \tag{12}$$

From the correspondence at points E($dw/dz = k \tan \alpha\pi$, $t = -\infty$), we get

$$A = \frac{ik \tan \alpha\pi}{J''} e^{i\pi\alpha} \tag{13a}$$

where

$$J'' = \int_1^{-\infty} (1 - t)^{-\frac{1}{2}-\alpha}(t - a)^{-1} dt$$
$$= (a - 1)^{-\frac{1}{2}-\alpha}\Gamma(\tfrac{1}{2} - \alpha)\Gamma(\tfrac{1}{2} + \alpha) \tag{13b}$$

Multiplying the expressions for dw/dz [Eq. (12)] and dz/dt [Eq. (7)], and integrating the resulting expression, we obtain the equation

$$w = MA \int_0^t \left[\zeta^{-\frac{1}{2}}(\zeta - 1)^{\alpha-1} \int_1^\zeta (\tau - 1)^{-\frac{1}{2}-\alpha}(\tau - a)^{-1} d\tau \right] d\zeta \tag{14}$$

Substituting the corresponding values at points E ($w = -kh + iq$, $t = -\infty$), we have, after some reduction,

$$q = \frac{kh}{JJ'' \cos \alpha\pi} \int_0^\infty \left[\zeta^{-\frac{1}{2}}(1 + \zeta)^{\alpha-1} \int_{-1}^\zeta (1 + \tau)^{-\frac{1}{2}-\alpha}(a + \tau)^{-1} d\tau \right] d\zeta \tag{15}$$

---

* See K. Pearson, "Tables of the Incomplete Beta-function," Cambridge University Press, London, 1956.

To evaluate Eq. (15), Mkhitarian substitutes $\tau = (au - 1)/(1 - u)$ for the interior integral and $\zeta = ap/(1 - p)$ for the exterior integral; then, expanding the resulting expression into power series and integrating term by term, he obtains for the discharge quantity

$$\frac{q}{kh} = \tan \alpha\pi - \frac{(a - 1)^{1/2 + \alpha}\Phi(\alpha,a)}{\sqrt{\pi}\,\Gamma(\alpha)\Gamma(1/2 - \alpha)} \tag{16}$$

where a is given implicitly by Eq. (11) and the function $\Phi(\alpha,a)$ can be evaluated from the following:

$$\Phi(\alpha,a) = a^{-\alpha}\left[C_0 \sum_{n=0}^{\infty} (-1)^n \frac{2}{2n + 1} A_n(a - 1)^n \right.$$

$$+ C_1 \frac{a - 1}{a} \sum_{n=0}^{\infty} \frac{(-1)^n 2^2 1!}{(2n + 3)(2n + 1)} A_n(a - 1)^n$$

$$\left. + C_2 \left(\frac{a - 1}{a}\right)^2 \sum_{n=0}^{\infty} (-1)^n \frac{2^3 2!}{(2n + 5)(2n + 3)(2n + 1)} A_n(a - 1)^n + \cdots \right]$$

where $C_0 = \dfrac{1}{1/2 + \alpha}$ $C_1 = \dfrac{1/2 + \alpha}{3/2 + \alpha}$ $C_2 = \dfrac{(1/2 + \alpha)(3/2 + \alpha)}{2!(5/2 + \alpha)}$

$$C_n = \frac{(1/2 + \alpha)(3/2 + \alpha)\cdots[(2n - 1)/2 + \alpha]}{n![(2n + 1)/2 + \alpha]}$$

$$A_0 = 1 \qquad A_1 = 1 - \alpha \qquad A_2 = \frac{(1 - \alpha)(2 - \alpha)}{2!}$$

$$A_n = \frac{(1 - \alpha)(2 - \alpha)\cdots(n - \alpha)}{n!}$$

It is immediately apparent from Eq. (16) that when tail water is absent ($a = 1$) the quantity of seepage for this case reduces to

$$q = kh \tan \alpha\pi \tag{17}$$

which is precisely the same result obtained from the assumption of horizontal flow [cf. Eq. (6) with $I = (\cot \alpha\pi)^{-1}$].

On the basis of the results given in Eqs. (6) and (17), from a practical point of view, it will suffice to calculate the quantity of seepage through cores of rockfill dams (without tail water) using the simplifying assumption of horizontal flow.

8-4. Rockfill Dams with Tail Water

From his solution [Eq. (16), Sec. 8-3] for the discharge through a triangular core with a vertical upstream slope (Fig. 8-4a), Mkhitarian computed q/kh as a function of the ratio of tail-water to headwater elevations (h_0/h) for downstream slopes of 2.5:1 ($\alpha = 0.121$) and 3:1 ($\alpha = 0.102$). The results are shown in Table 8-2 and are plotted in Fig. 8-6. In this figure q/q_{max} is the ratio of the discharges with and without tail water.

Table 8-2*

2.5:1, $\alpha = 0.1211$

$\dfrac{h_0}{h}$	0.870	0.769	0.500	0.312	0.200	0.143
$\dfrac{q}{kh}$	0.241	0.318	0.390	0.399	0.400	0.400

3.0:1, $\alpha = 0.1024$

$\dfrac{h_0}{h}$	0.909	0.860	0.667	0.556	0.400	0.294	0.192
$\dfrac{q}{kh}$	0.161	0.261	0.305	0.322	0.331	0.332	0.333

* From A. M. Mkhitarian, Computation of Seepage through Earth Dams, *Inzhenernil Sbornik*, vol. 14, 1952.

The exact solution for the discharge quantity for a section with vertical slopes, including the effects of tail water, was shown by Charny [18] to be given by Dupuit's formula [Eq. (5), Sec. 2-2]

$$q = k \frac{h^2 - h_0^2}{2L} \tag{1}*$$

A plot of this expression as a function of tail-water height and width of section is given in Fig. 8-5a. The elevation of the exit point of the free surface as a function of the characteristic dimensions of the section was obtained by Polubarinova—Kochina [116] and is given in Fig. 8-5b. A plot of the variation in discharge with tail-water for this case is also shown in

* Experimental verification was provided by Wyckoff and Reed [161] who used an electrical analogue (Sec. 6-2) in which the shape of the free surface was determined by trial and error.

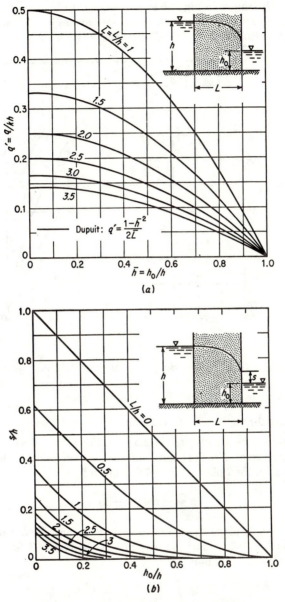

FIG. 8-5. (*After Polubarinova-Kochina* [116].)

Fig. 8-6. Also shown in Fig. 8-6 is the plot of the discharge as a function of tail-water variation for a core with base angles of 45°, as given by Polubarinova-Kochina. Since the upstream slope has only a small

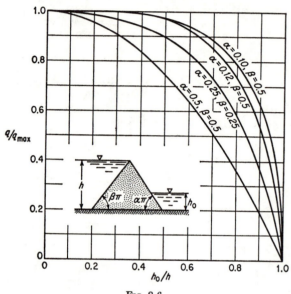

Fig. 8-6

influence on the discharge (cf. Sec. 3-2), for practical purposes the reduction in the discharge with tail-water variations for any configuration may be obtained by interpolation (of $\alpha\pi$) from the figure.

8-5. Seepage through an Earth Dam on an Inclined Impervious Base

The section to be investigated is shown in Fig. 8-7a. We shall assume that the free surface CD_∞ lies below point E. The mapping of the w plane (Fig. 8-7b) onto the t plane (Fig. 8-7c) is given by

$$w = \frac{iq}{\pi} \cos^{-1} t - kh \tag{1}$$

The hodograph plane (assuming $\alpha + \beta < \frac{1}{2}$) and its inverse dz/dw are given in Fig. 8-7d. For the mapping of the region dz/dw onto the t plane (Fig. 8-7c) we obtain

$$\frac{dz}{dw} = M \int_0^t \frac{(t-1)^{\alpha+\beta-\frac{3}{2}}}{(t+1)^{\beta+\frac{1}{2}}} dt + N \tag{2}$$

To determine the constants M and N, we note that at point C in the hodograph plane $u = k \sin \pi(\frac{1}{2} - \beta) \cos \pi(\frac{1}{2} - \beta)$ and $v = k \sin^2 \pi(\frac{1}{2} - \beta)$

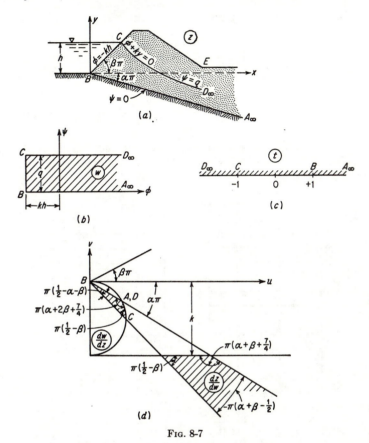

FIG. 8-7

whereas in the t plane, $t = -1$. Substituting into Eq. (2), after some elaboration we get

$$\frac{dz}{dw} = - \frac{ie^{\pi\beta i} \cos \pi(\alpha + \beta)}{Jk \cos \pi\alpha \cos \pi\beta} \int_{-1}^{t} (1 - t)^{\alpha+\beta-\frac{3}{2}}(1 + t)^{\frac{1}{2}-\beta} \, dt - \frac{ie^{\pi\beta i}}{k \cos \pi\beta} \quad (3)$$

where $$J = 2^{\alpha-1} \frac{\Gamma(\frac{1}{2})\Gamma(\frac{1}{2} - \alpha)}{\Gamma(\frac{3}{2} - \beta - \alpha)} \quad (4)$$

Multiplying dz/dw [Eq. (3)] by dw/dt [first differentiating Eq. (1) with respect to t] and expanding the resulting expression in a power series, after term-by-term integration we have for the quantity of seepage

$$q = \frac{kh \cot \pi\beta}{1 + [\cot \pi(\alpha + \beta)/(J\pi \sin \alpha\pi)](J_1 + J_2)} \quad (5)$$

where J_1 and J_2 are given by the series

$$J_1 = \frac{1}{\frac{1}{2} - \beta}\left[\frac{\pi}{2} + 2^{\frac{1}{2}-\beta}\left(\frac{0.5^{1-\beta}}{1-\beta} + \frac{1}{2}\frac{0.5^{2-\beta}}{2-\beta} + \frac{1}{3}\frac{0.5^{3-\beta}}{3-\beta} + \cdots\right)\right]$$

$$J_2 = -\frac{1}{\alpha + \beta - \frac{1}{2}}\left[\frac{\pi}{2} + 2^{\alpha+\beta-\frac{1}{2}}\left(\frac{0.5^{\alpha+\beta+1}}{\alpha+\beta+1}\right.\right. \qquad (6)$$
$$\left.\left. + \frac{1}{2}\frac{0.5^{\alpha+\beta+2}}{\alpha+\beta+2} + \cdots\right)\right]$$

The above solution was developed by Mkhitarian [96] in 1953. A plot of J_1 as a function of β is given in Fig. 8-8a, and of J_2 as a function of $\alpha + \beta$ in Fig. 8-8b.

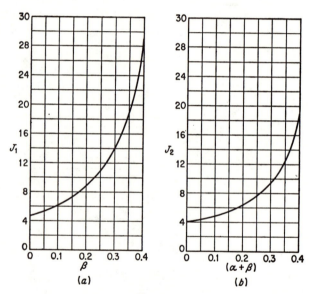

Fig. 8-8

8-6. Earth Dam on Impervious Base with Toe Filter

The solution for the seepage characteristics of the dam section shown in Fig. 8-9a was obtained by Numerov [104] in 1942. Although the derivation of the solution is quite lengthy the methodology is of sufficient value to receive full development in this text. In addition graphs are provided which render the solution amenable to direct and simple engineering use.

If we neglect the surface of seepage (more will be said concerning this later), the w plane is as given in Fig. 8-9b. Selecting an auxiliary t plane

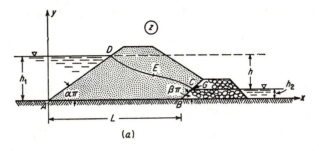

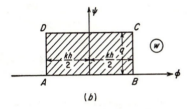

Fig. 8-9

as shown in Fig. 8-9c, we obtain for the mapping of the w plane

$$t = m \operatorname{sn} \frac{2Kw}{kh} \tag{1}$$

and for the discharge

$$q = \frac{khK'}{2K} \tag{2}$$

In particular, in the intervals $m < t < 1$ (plus) and $-1 < t < -m$ (minus), we have

$$w = \pm \frac{kh}{2} + iq \left[1 - \frac{F(m',\lambda)}{K'} \right] \tag{3}$$

where $\lambda = \sin^{-1} (\sqrt{1 - t^2}/m')$.

To obtain the relationship between the z plane and the w plane, Numerov introduced the auxiliary function

$$Z = X + iY = z + \frac{1}{2kq} \left[\frac{k(h_1 + h_2)}{2} - w \right]^2 \tag{4}*$$

* See Prob. 7 of this chapter.

Let us investigate the nature of this function on the boundaries of the flow region.

Along AB, where $\psi = 0$ and $y = 0$,

$$Y = 0 \tag{5a}$$

Along the line BC, the equation of which can be written as

$$x \sin \beta\pi - y \cos \beta\pi = L \sin \beta\pi$$

and along which $\phi = kh/2$, we have

$$X \sin \beta\pi - Y \cos \beta\pi = \left(L + \frac{kh_2{}^2}{2q}\right) \sin \beta\pi + \frac{h_2 \cos \beta\pi}{q} \psi - \frac{\sin \beta\pi}{2kq} \psi^2 \tag{5b}$$

Along the free surface CD, where $\phi = k(h_1 + h_2)/2 - ky$ and $\psi = q$,

$$Y = 0 \tag{5c}$$

Finally, along the upstream slope DA, the equation of which is $x \sin \alpha\pi - y \cos \alpha\pi = 0$, and along which $\phi = -kh/2$, we have

$$X \sin \alpha\pi - Y \cos \alpha\pi = \frac{kh_1{}^2 \sin \alpha\pi}{2q} + \frac{h_1 \cos \alpha\pi}{q} \psi - \frac{\sin \alpha\pi}{2kq} \psi^2 \tag{5d}$$

From Eqs. (5) we see that along the various segments of the contour of the w plane the unknown analytic function $Z = X + iY$ can be expressed as $aX + bY = c$, where a and b are constants and c is the function on the boundary. Thus the problem resolves itself into seeking the function $Z(w)$ interior to w. This is the *Reimann-Hilbert* problem. Noting that along those boundaries of the flow domain where $Y \neq 0$ [Eqs. (5b) and (5d)], $\phi = \pm kh/2$, we find from Eq. (3) for the intervals $m \leqq r \leqq 1$, $-1 \leqq r \leqq -m$, and $r = \text{Re}(t)$ that

$$\psi = q \left[1 - \frac{F(m',\lambda)}{K'} \right] \tag{6}$$

where $\lambda = \sin^{-1}(\sqrt{1 - r^2}/m')$. Substituting Eq. (6) for ψ into Eqs. (5b) and (5d), we obtain after some simplification:

for $m < r < 1$:

$$X \sin \beta\pi - Y \cos \beta\pi = b_0 - \frac{b_1}{K'} F(m',\lambda) - \frac{b_2}{K'^2} F^2(m',\lambda) = \gamma_1$$

for $-1 < r < -m$:

$$X \sin \alpha\pi - Y \cos \alpha\pi = a_0 - \frac{a_1}{K'} F(m',\lambda) - \frac{a_2}{K'^2} F^2(m',\lambda) = \gamma_2 \tag{7a}$$

for $|r| < m$ and $|r| > 1$:

$$Y = 0$$

where $$b_0 = L \sin \beta\pi + h_2 \cos \beta\pi + \frac{kh_2{}^2 \sin \beta\pi}{2q} - \frac{q \sin \beta\pi}{2k}$$

$$b_1 = h_2 \cos \beta\pi - \frac{q \sin \beta\pi}{k}$$

$$b_2 = \frac{q \sin \beta\pi}{2k}$$

$$a_0 = h_1 \cos \alpha\pi + \frac{kh_1{}^2 \sin \alpha\pi}{2q} - \frac{q \sin \alpha\pi}{2k}$$ $\qquad(7b)$

$$a_1 = h_1 \cos \alpha\pi - \frac{q \sin \alpha\pi}{k}$$

$$a_2 = \frac{q \sin \alpha\pi}{2k}$$

At this point, Numerov introduced the function

$$T(t) = i(m - t)^\beta (m + t)^{1-\alpha}(1 - t)^{1-\beta}(1 + t)^\alpha \qquad (8a)$$

Examining this function, we see that it yields imaginary values for real values of t [$r = \text{Re } (t)$] in the interval $-m < r < m$ ($m \leqq 1$). Therefore its argument will be $\pi/2$ for $0 < r < m$, and on the various sections of the real axis of t we will have the following conditions:

for $|r| < m$:
$$T(r) = i(m - r)^\beta (m + r)^{1-\alpha}(1 - r)^{1-\beta}(1 + r)^\alpha$$
for $m < r < 1$:
$$T(r) = (r - m)^\beta (m + r)^{1-\alpha}(1 - r)^{1-\beta}(1 + r)^\alpha e^{\pi i(\frac{1}{2}-\beta)}$$
for $-1 < r < -m$:
$$T(r) = (m - r)^\beta (-m - r)^{1-\alpha}(1 - r)^{1-\beta}(1 + r)^\alpha e^{\pi i(\frac{1}{2}-\alpha)}$$ $\qquad (8b)$
for $-\infty < r < -1$:
$$T(r) = -i(m - r)^\beta (-m - r)^{1-\alpha}(1 - r)^{1-\beta}(-1 - r)^\alpha$$
for $1 < r < \infty$:
$$T(r) = -i(r - m)^\beta (m + r)^{1-\alpha}(r - 1)^{1-\beta}(1 + r)^\alpha$$

Noting that $\text{Re } [e^{\pi i(\frac{1}{2}-\alpha)}Z] = X \sin \alpha\pi - Y \cos \alpha\pi$, we see that the product

$$T(t)Z(t) \qquad (9)$$

will be imaginary for $|r| < m$ and $|r| > 1$; that is, the real part of the product is zero for these sections of the real axis. Within the intervals $m < r < 1$ and $-1 < r < -m$ the real part of Eq. (9) is known from Eqs. (5) and (8).

Let us now investigate the nature of the product $T(t)Z(t)$ in the vicinity of $t = \infty$. As $Z(t)$ is an analytic function bounded in the half plane t, it can be expanded into the series

$$Z(t) = \sum_{n=0}^{\infty} \frac{c_n}{t^n} \tag{10a}$$

where, since for $|r| > 1$ the imaginary part of Z is zero ($Y = 0$), the constants c_n must be real. For the function $T(t)$ in the vicinity of infinity, from Eqs. (8b) we obtain

$$T(t) = -it^2 \left(1 - \frac{m}{t}\right)^\beta \left(1 + \frac{m}{t}\right)^{1-\alpha} \left(1 - \frac{1}{t}\right)^{1-\beta} \left(1 + \frac{1}{t}\right)^\alpha$$

$$= -it^2 \left(1 + \sum_{n=1}^{\infty} \frac{d_n}{t^n}\right) \tag{10b}$$

where the coefficients d_n are real. Hence from Eqs. (10) in the vicinity of infinity we have

$$T(t)Z(t) = -it^2 \left(1 + \frac{d_1}{t} + \frac{d_2}{t^2} + \cdots\right) \left(c_0 + \frac{c_1}{t} + \frac{c_2}{t^2} + \cdots\right)$$

$$= -ic_0 t^2 - i(c_1 + c_0 d_1)t - i(c_0 + c_1 d_1 + c_0 d_2) + \cdots \tag{10c}$$

Now the new function

$$W(t) = T(t)Z(t) + ic_0 t^2 + i(c_1 + c_0 d_1)t \tag{11a}$$

is bounded at infinity. In addition, it is analytic in the entire half plane t (including $t = \infty$) and single-valued and continuous up to the real axis, on which the following conditions hold:

$$f(r) = \operatorname{Re} W(r) = \begin{cases} \gamma_1(r)\delta_1(r) & m < r < 1 \\ -\gamma_2(r)\delta_2(r) & -1 < r < -m \\ 0 & |r| < m \text{ and } |r| > 1 \end{cases} \tag{11b}$$

where $\gamma_1(r)$ and $\gamma_2(r)$ are as given in Eq. (7a), and

$$\delta_1(r) = (r - m)^\beta (m + r)^{1-\alpha}(1 - r)^{1-\beta}(1 + r)^\alpha$$
$$\delta_2(r) = (m - r)^\beta (-m - r)^{1-\alpha}(1 - r)^{1-\beta}(1 + r)^\alpha$$

Now, instead of linear combinations of two functions X and Y, only one function needs to be considered: the real part of W. (This is the Dirichlet problem for a half plane; cf. Ref. 20.) Consequently, we can write for $W(t)$

$$W(t) = -\frac{i}{\pi} \int_{-1}^{1} \frac{f(\tau)\, d\tau}{\tau - t} + iC \tag{12}$$

where C is an arbitrary real constant. Substituting for $W(t)$ from Eqs. $(8a)$ and $(11a)$, we get (dividing through by i)

$$(c_0 d_1 + c_1)t + c_0 t^2 + (m - t)^\beta (m + t)^{1-\alpha}(1 - t)^{1-\beta}(1 + t)^\alpha Z(t)$$
$$= -\frac{1}{\pi} \int_{-1}^{1} \frac{f(\tau)\, d\tau}{\tau - t} + C \quad (13)$$

To obtain the constants c_0, c_1, d, and C, we substitute into Eq. (13) successively $t = \pm m$ and $t = \pm 1$. Substituting $t = m$ and subtracting the resulting expression from Eq. (13), we find

$$c_0 d_1 + c_1 + c_0(m + t) - (m - t)^{\beta-1}(m + t)^{1-\alpha}(1 - t)^{1-\beta}(1 + t)^\alpha Z(t)$$
$$= -\frac{1}{\pi} \int_{-1}^{1} \frac{f(\tau)\, d\tau}{(\tau - m)(\tau - t)}$$

Similarly, substituting $t = -m$ into the preceding expression and subtracting, we have

$$c_0 - (m - t)^{\beta-1}(m + t)^{-\alpha}(1 - t)^{1-\beta}(1 + t)^\alpha Z(t)$$
$$= -\frac{1}{\pi} \int_{-1}^{1} \frac{f(\tau)\, d\tau}{(\tau - m)(\tau + m)(\tau - t)}$$

Taking $t = 1$ and subtracting, we obtain

$$(m - t)^{\beta-1}(m + t)^{-\alpha}(1 - t)^{-\beta}(1 + t)^\alpha Z(t)$$
$$= -\frac{1}{\pi} \int_{-1}^{1} \frac{f(\tau)\, d\tau}{(\tau^2 - m^2)(\tau - 1)(\tau - t)} \quad (14)$$

Finally, taking $t = -1$, the left part of Eq. (14) becomes zero and

$$\int_{-1}^{1} \frac{f(\tau)\, d\tau}{(m^2 - \tau^2)(1 - \tau^2)} = 0 \quad (15)$$

Subtracting Eq. (15) from Eq. (14) and solving for $Z(t)$, the resulting expression is given in symmetrical form as

$$Z(t) = -\frac{1}{\pi}(m - t)^{1-\beta}(m + t)^\alpha(1 - t)^\beta(1 + t)^{1-\alpha}$$
$$\int_{-1}^{1} \frac{f(\tau)\, d\tau}{(m^2 - \tau^2)(1 - \tau^2)(\tau - t)} \quad (16a)$$

which, making use of Eq. (4), becomes

$$z = -\frac{1}{2kq}\left[\frac{k(h_1 + h_2)}{2} - w\right]^2 - \frac{T_1(t)}{\pi}\int_{-1}^{1} \frac{f(\tau)\, d\tau}{(m^2 - \tau^2)(1 - \tau^2)(\tau - t)} \quad (16b)$$

where $\qquad T_1(t) = (m - t)^{1-\beta}(m + t)^\alpha(1 - t)^\beta(1 + t)^{1-\alpha}$

Although the above work yields the complete solution for the seepage characteristics through the dam [that is, the modulus m is given by

Eq. (15), the quantity by Eq. (2), and the characteristic function by Eq. (16b)], the resulting equations are not amenable to direct engineering use. To obviate this difficulty, Numerov noted that in most practical cases, the ratio L/h in earth dams is generally large, often of the order of $L/h > 10$. With L/h large, g/kh will be small, the modulus m will be near unity, $m' \approx 0$, and $F(m',\lambda) \approx \lambda$. Before applying this approximation, we shall reduce Eq. (15) to a more compatible form. To do this, we first substitute into this equation Eqs. (7a) and (11b), obtaining

$$\int_{-1}^{-m} \left(a_0 - \frac{a_1}{K'} F - \frac{a_2}{K'^2} F^2 \right) \frac{\delta_2(\tau)\, d\tau}{(\tau^2 - m^2)(1 - \tau^2)}$$
$$- \int_{m}^{1} \left(b_0 - \frac{b_1}{K'} F - \frac{b_2}{K'^2} F^2 \right) \frac{\delta_1(\tau)\, d\tau}{(\tau^2 - m^2)(1 - \tau^2)} = 0 \quad (17)$$

where $F = F(m',\lambda)$ and δ_1 and δ_2 are as defined in Eq. (11b). Then, introducing the integrals:

$$J_0 = \int_{-1}^{-m} \frac{\delta_2(\tau)\, d\tau}{(\tau^2 - m^2)(1 - \tau^2)} = \int_{m}^{1} \Phi(\tau)\, d\tau \qquad J'_0 = \int_{m}^{1} \Phi_1(\tau)\, d\tau$$

$$J_1 = \int_{m}^{1} F\Phi(\tau)\, d\tau \qquad\qquad\qquad\qquad J'_1 = \int_{m}^{1} F\Phi_1(\tau)\, d\tau$$

$$J_2 = \int_{m}^{1} F^2\Phi(\tau)\, d\tau \qquad\qquad\qquad\quad J'_2 = \int_{m}^{1} F^2\Phi_1(\tau)\, d\tau$$

where
$$\Phi(\tau) = (m + \tau)^{\beta-1}(\tau - m)^{-\alpha}(1 + \tau)^{-\beta}(1 - \tau)^{\alpha-1}$$
$$\Phi_1(\tau) = (\tau - m)^{\beta-1}(\tau + m)^{-\alpha}(1 - \tau)^{-\beta}(1 + \tau)^{\alpha-1}$$

we find Eq. (17) can be written as

$$a_0 J_0 - \frac{a_1}{K'} J_1 - \frac{a_2}{K'^2} J_2 - b_0 J'_0 + \frac{b_1}{K'} J'_1 + \frac{b_2}{K'^2} J'_2 = 0 \qquad (18)$$

Now, making the above stated assumptions* and considering the integral J_0, we have

$$J_0 \approx \tfrac{1}{2} \int_{m}^{1} (\tau - m)^{-\alpha}(1 - \tau)^{\alpha-1}\, d\tau$$

which, substituting $1 - \tau = (1 - m)x$, yields the well-known beta function

$$J_0 \approx \frac{1}{2} \int_{0}^{1} x^{\alpha-1}(1 - x)^{-\alpha}\, dx = \frac{1}{2} B(\alpha, 1 - \alpha) = \frac{\pi}{2 \sin \alpha\pi}$$

By exactly the same procedure we find

$$J'_0 \approx \frac{\pi}{2 \sin \beta\pi}$$

* In particular, it is assumed that $(m + \tau)^{\beta-1} \approx (1 + \tau)^{\beta-1}$, $(1 + \tau)^{-1} = (\tfrac{1}{2})$ $[1 - (\tau - 1)/2 + \cdots] \approx \tfrac{1}{2}$, and $F(m',\lambda) \approx \lambda = \sin^{-1}(\sqrt{1 - \tau^2}/m')$.

For the integral J_1, making the same assumptions as for J_0, we get

$$J_1 \approx \frac{1}{2} \int_m^1 (\tau - m)^{-\alpha}(1 - \tau)^{\alpha-1} \sin^{-1} \frac{\sqrt{1 - \tau^2}}{m'} \, d\tau$$

If

$$\sin^{-1} \frac{\sqrt{1 - \tau^2}}{m'} = \frac{\pi x}{2}$$

$$\tau = \sqrt{1 - m'^2 \sin^2 \frac{\pi x}{2}} \approx 1 - \frac{1}{2} m'^2 \sin^2 \frac{\pi x}{2}$$

$$d\tau \approx -\frac{\pi}{2} m'^2 \sin \frac{\pi x}{2} \cos \frac{\pi x}{2} \, dx$$

$$1 - \tau \approx \frac{1}{2} m'^2 \sin^2 \frac{\pi x}{2}$$

$$\tau - m \approx (1 - m)\left[1 - \frac{1}{2}(1 + m)\sin^2 \frac{\pi x}{2}\right] \approx (1 - m)\cos^2 \frac{\pi x}{2}$$

this yields

$$J_1 \approx \frac{\pi^2}{4} \int_0^1 x \cot^{1-2\alpha} \frac{\pi x}{2} \, dx = \frac{\pi^2}{4} f_1(\alpha)$$

By the same procedure we find

$$J_2 \approx \frac{\pi^3}{8} \int_0^1 x^2 \cot^{1-2\alpha} \frac{\pi x}{2} \, dx = \frac{\pi^3}{4} f_2(\alpha)$$

$$J_1' \approx \frac{1}{2} \int_m^1 (\tau - m)^{\beta-1}(1 - \tau)^{-\beta} \sin^{-1} \frac{\sqrt{1 - \tau^2}}{m'} \, d\tau$$

$$= \frac{\pi^2}{4} \int_0^1 x \cot^{2\beta-1} \frac{\pi x}{2} \, dx = \frac{\pi^2}{4} f_1(1 - \beta)$$

$$J_2' \approx \frac{\pi^3}{8} \int_0^1 x^2 \cot^{2\beta-1} \frac{\pi x}{2} \, dx = \frac{\pi^3}{4} f_2(1 - \beta)$$

Noting that for m' close to zero $K' = \pi/2$, upon substitution of the above expressions for $J_0, J_1, \ldots$, and the constants $a_0, a_1, \ldots, b_0, b_1, \ldots$, [from Eq. (7b)] into Eq. (18), we obtain after some simplification

$$L \approx \frac{k(h_1^2 - h_2^2)}{2q} + h_1[\cot \alpha\pi - f_1(\alpha) \cos \alpha\pi]$$

$$- h_2[\cot \beta\pi - f_1(1 - \beta) \cos \beta\pi] - \frac{q}{k}[-f_2(\alpha) \sin \alpha\pi + f_2(1 - \beta) \sin \beta\pi]$$

where $\quad f_3(\alpha) = f_1(\alpha) - f_2(\alpha) = \int_0^1 x\left(1 - \frac{x}{2}\right) \cot^{1-2\alpha} \frac{\pi x}{2} \, dx$

To simplify the last expression, we note that the substitution of $x = 1 - y$ yields

$$f_1(1 - \beta) = \int_0^1 \cot^{1-2\beta} \frac{\pi y}{2}\, dy - \int_0^1 y \cot^{1-2\beta} \frac{\pi y}{2}\, dy = \frac{1}{\sin \beta\pi} - f_1(\beta)$$

$$f_3(1 - \beta) = \frac{1}{2} \int_0^1 (1 - y)^2 \cot^{1-2\beta} \frac{\pi y}{2}\, dy = \frac{1}{2 \sin \beta\pi} - f_1(\beta) + f_3(\beta)$$

$$f_2(1 - \beta) = \int_0^1 (1 - y)\left(\frac{1}{2} + \frac{y}{2}\right) \cot^{1-2\beta} \frac{\pi y}{2}\, dy = \frac{1}{2 \sin \beta\pi} - \frac{1}{2} f_3(\beta)$$

whence
$$L \approx \frac{k(h_1{}^2 - h_2{}^2)}{2q} + h_1[\cot \alpha\pi - f_1(\alpha) \cos \alpha\pi]$$
$$- h_2 f_1(\beta) \cos \beta\pi - \frac{q}{k}\left[\frac{1}{2} - f_2(\alpha) \sin \alpha\pi - f_3(\beta) \sin \beta\pi\right] \quad (19)$$

Equation (19) yields L as a function of q/k (or q/k as a function of L) with great precision. The functions f_1, f_2, and f_3 are given in Fig. 8-10.

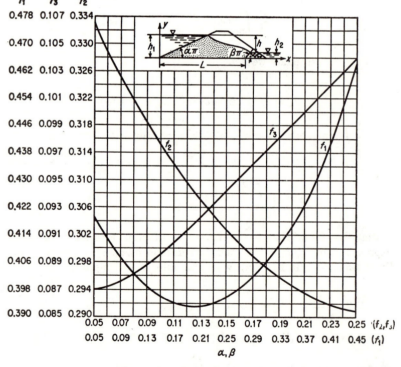

Fig. 8-10. (*After Shankin* [132].)

In the case of a rectangular dam with a horizontal underdrain, $\alpha = \frac{1}{2}$, $\beta = 0$, $h_1 = h$, $h_2 = 0$, and Eq. (19) reduces to

$$q \approx \frac{kh^2}{L + \sqrt{L^2 + h^2/3}} \qquad (20)$$

For the equation of the free surface, substituting into the characteristic function [Eq. (16b)] the complex potential along this surface $[w = \phi + iq = k(h_1 + h_2)/2 - ky + iq]$, we obtain

$$x = -\frac{ky^2}{2q} + \frac{q}{2k} + \frac{1}{\pi}(m - r)^{1-\beta}(-m - r)^{\alpha}(1 - r)^{\beta}(-1 - r)^{1-\alpha}$$
$$\left\{ \int_{-1}^{-m} \frac{[a_0 - (a_1/K')F - (a_2/K'^2)F^2]\,d\tau}{(m - \tau)^{1-\beta}(-m - \tau)^{\alpha}(1 + \tau)^{1-\alpha}(\tau - r)(1 - \tau)^{\beta}} \right.$$
$$\left. - \int_{1}^{m} \frac{[b_0 - (b_1/K')F - (b_2/K'^2)F^2]\,d\tau}{(\tau - m)^{1-\beta}(m + \tau)^{\alpha}(1 - \tau)^{\beta}(1 + \tau)^{1-\alpha}(\tau - r)} \right\} \qquad (21)$$

Considering the left side of the free surface, where $-\infty < r < -1$ (between D and E in Fig. 8-9a), we shall again take $m \approx 1$, $K' \approx \pi/2$, and $F(m',\lambda) \approx \lambda$; hence

$$x \approx -\frac{ky^2}{2q} + \frac{q}{2k} + \frac{1}{\pi}(1 - r)(-m - r)^{\alpha}(-1 - r)^{1-\alpha}$$
$$\left\{ \frac{1}{2} \int_{-1}^{-m} \frac{[a_0 - (2a_1/\pi)F - (4a_2/\pi^2)F^2]\,d\tau}{(-m - \tau)^{\alpha}(1 + \tau)^{1-\alpha}(\tau - r)} \right.$$
$$\left. - \frac{1}{2} \int_{m}^{1} \frac{[b_0 - (2b_1/\pi)F - (4b_2/\pi^2)F^2]\,d\tau}{(\tau - m)^{1-\beta}(1 - \tau)^{\beta}(\tau - r)} \right\} \qquad (22)^*$$

We note that in the above integrals the term $\tau - r$ can be replaced with sufficient accuracy by $1 - r$; that is, for $m \approx 1$, $t \approx 1$. Making the same changes in Eq. (17), we obtain

$$\int_{-1}^{-m} \frac{a_0 - (2a_1/\pi)F - (4a_2/\pi^2)F^2}{(-m - \tau)^{\alpha}(1 + \tau)^{1-\alpha}}\,d\tau \approx \int_{m}^{1} \frac{b_0 - (2b_1/\pi)F - (4b_2/\pi^2)F^2}{(\tau - m)^{1-\beta}(1 - \tau)^{\beta}}\,d\tau$$

Multiplying both sides of this equality by $(-\tau - r)^{\alpha}(-1 - r)^{1-\alpha}/\pi$ and subtracting the resulting expression from Eq. (22), after noting that

---

* See previous footnote.

$(1 - r)/(\tau - r) - 1 = (1 - \tau)/(\tau - r)$, we obtain

$$x \approx -\frac{ky^2}{2q} + \frac{q}{2k}$$
$$+ \frac{1}{\pi}(-m - r)^\alpha(-1 - r)^{1-\alpha} \int_{-1}^{-m} \frac{a_0 - (2a_1/\pi)F - (4a_2/\pi^2)F^2}{(-m - \tau)^\alpha(1 + \tau)^{1-\alpha}(\tau - r)} d\tau \quad (23a)$$

Now, making the change of variable

$$\tau = \frac{-r - m + (1 - m)r \sin^2 (\pi t/2)}{-r - m - (1 - m) \sin^2 (\pi t/2)} \quad (23b)$$

in Eq. (23a), we obtain

$$x \approx -\frac{ky^2}{2q} + \frac{q}{2k} + \int_0^1 \cot^{1-2\alpha} \frac{\pi t}{2} \left(a_0 - \frac{2a_1}{\pi} F - \frac{4a_2}{\pi^2} F^2\right) dt \quad (23c)$$

Taking $\qquad F \approx \sin^{-1} \sqrt{\dfrac{1 - \tau^2}{1 - m^2}} \approx \sin^{-1} \sqrt{\dfrac{1 + \tau}{1 - m}}$

as $(1 - \tau)/(1 + m) \approx \frac{2}{2} = 1$, and changing $\sin^{-1} \chi$ into $\tan^{-1} \chi/\sqrt{1 - \chi^2}$, we get $F \approx \tan^{-1} \sqrt{(1 + \tau)/(-\tau - m)}$. Hence from Eq. (23b) we find

$$F \approx \tan^{-1} \left(\tan \frac{\pi t}{2} \sqrt{\frac{-1 - r}{-m - r}}\right) \quad (24)$$

From Eq. (1) for the interval $-\infty < r < -1, r = m$ sn $[2K(\phi + iq)/kh]$, which, on the basis of Eq. (2), can be written as

$$r = m \text{ sn} \left(\frac{K'\phi}{q} + iK', m\right)$$

For the left side of the free surface where y is close to h_1, we can write

$$\frac{\phi}{q} = \frac{k(h_1 + h_2)}{2q} - \frac{ky}{q} = \frac{k(h_1 - y)}{q} - \frac{kh}{2q} = \frac{k(h_1 - y)}{q} - \frac{K}{K'}$$

Now, noting (cf. Appendix B) that m sn $(\chi + iK', m) = 1/\text{sn} (\chi, m)$, and sn $(\chi - K, m) = -$ cn $(\chi, m)/\text{dn} (\chi, m)$, we obtain

$$r = \frac{1}{\text{sn} (K'\phi/q, m)} = \frac{\text{dn} [K'k(h_1 - y)/q, m]}{\text{cn} [K'k(h_1 - y)/q, m]} \quad (25)$$

whereas for $m \approx 1$, $(r + 1)/(r + m) \approx 1$, and

$$u = \sqrt{\frac{-1 - r}{-m - r}} \approx \sqrt{\frac{r^2 - 1}{r^2 - m^2}}$$

whence, substituting r from Eq. (25), we find

$$u = \sqrt{\frac{\mathrm{dn}^2 \chi - \mathrm{cn}^2 \chi}{\mathrm{dn}^2 \chi - m^2 \mathrm{cn}^2 \chi}} = \mathrm{sn}\left[\frac{K'k(h_1 - y)}{q}, m\right] \approx \tanh \frac{\pi k(h_1 - y)}{2q} \quad (26)$$

Consequently, Eq. (24) can be expressed as

$$F \approx \tan^{-1}\left(u \tan \frac{\pi t}{2}\right)$$

Finally, making use of the above simplifications and substituting the equivalent expressions for a_0, a_1, and a_2 from Eqs. (7b), we obtain for the left side of the free surface

$$x \approx \frac{k(h_1^2 - y^2)}{2q} + h_1[\cot \alpha \pi - F_1(u,\alpha)] + \frac{q}{k} \tan \alpha \pi\, F_2(u,\alpha) \quad (27a)$$

where
$$u = \tanh \frac{k\pi(h_1 - y)}{2q}$$

$$F_1(u,\alpha) = \frac{2 \cos \alpha \pi}{\pi} \int_0^1 \cot^{1-2\alpha} \frac{\pi t}{2} \tan^{-1}\left(u \tan \frac{\pi t}{2}\right) dt$$

$$F_2(u,\alpha) = \frac{2 \cos \alpha \pi}{\pi} \int_0^1 \cot^{1-2\alpha} \frac{\pi t}{2} \tan^{-1}\left(u \tan \frac{\pi t}{2}\right)$$
$$\left[1 - \frac{1}{\pi} \tan^{-1}\left(u \tan \frac{\pi t}{2}\right)\right] dt$$

By the same procedure the equation of the right part of the free surface is

$$x \approx L + \frac{k(h_2^2 - y^2)}{2q} + h_2 F_3(u,\beta) + \frac{q}{k} F_4(u,\beta) \quad (27b)$$

where
$$u = \tanh \frac{k\pi(y - h_2)}{2q}$$

$$F_3(u,\beta) = \frac{2 \cos \beta \pi}{\pi} \int_0^1 \tan^{1-2\beta} \frac{\pi t}{2} \cot^{-1}\left(u \tan \frac{\pi t}{2}\right) dt$$

$$F_4(u,\beta) = \frac{1}{2} - \frac{2}{\pi^2} \sin \beta \pi \int_0^1 \tan^{1-2\beta} \frac{\pi t}{2} \left[\cot^{-1}\left(u \tan \frac{\pi t}{2}\right)\right]^2 dt$$

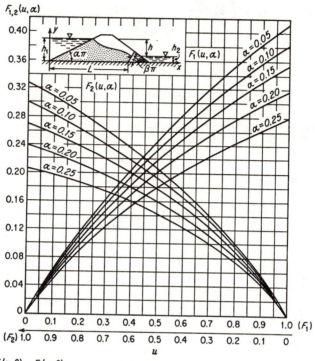

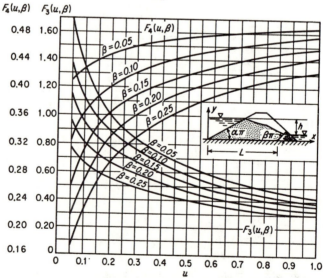

Fig. 8-11. (*After Shankin* [132].)

The functions $F_1(u,\alpha)$, $F_2(u,\alpha)$, $F_3(u,\beta)$, and $F_4(u,\beta)$ were tabulated by Shankin [132] and are shown in Fig. 8-11.   Hence, all the required information to effect a solution has been reduced to simple graphical form; that is, q/k can be determined from Eq. (19), and then the equation of the free surface can be obtained from Eqs. (27).

For a dam with a horizontal underdrain $\beta = 0$ (Fig. 8-12a), the shape of the free surface will be given by [Eq. (27a)],

$$x = k\frac{h^2 - y^2}{2q} + h[\cot \alpha\pi - F_1(u,\alpha)] + \frac{q}{k}\tan \alpha\pi F_2(u,\alpha) \qquad (28)$$

where $u = \tanh[k\pi(h-y)/2q]$.   The free surface for $\alpha = 0.25, L = 3.16h$, and $q = 0.2kh$ is shown in Fig. 8-12b.

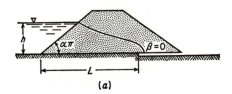

(a)

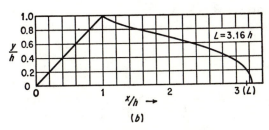

(b)

FIG. 8-12. *(After Numerov [104].)*

For a dam with a vertical upstream slope ($\alpha = \frac{1}{2}$) and an horizontal underdrain ($\beta = 0$), the equation of the free surface is

$$x = k\frac{h^2 - y^2}{2q} + \frac{q}{k}F\left(\tanh k\pi\frac{h-y}{2q}\right) \qquad (29)$$

where

$$u = \tanh\left(k\pi\frac{h-y}{2q}\right)$$

$$F(u) = \frac{2}{\pi}\int_0^1 \tan^{-1}\left(u\tan\frac{\pi t}{2}\right)\left[1 - \frac{1}{\pi}\tan^{-1}\left(u\tan\frac{\pi t}{2}\right)\right]dt$$

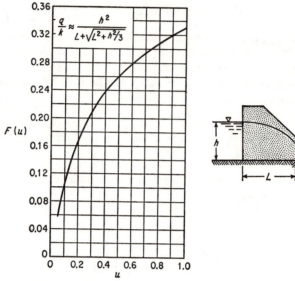

FIG. 8-13. (*After Aravin and Numerov* [2].)

A plot of $F(u)$ as a function of u is given in Fig. 8-13. q/k for this case is given by Eq. (20).

Let us now investigate the question of the development of the surface of seepage along CB of Fig. 8-9a.   Differentiating Eq. (27b) with respect to y,

$$\frac{dx}{dy} = -\frac{ky}{q} + \left\{ \left(\sin \beta \pi - \frac{kh_2}{q} \cos \beta \pi \right) \int_0^1 \frac{\tan^{2-2\beta}(\pi t/2)\, dt}{1 - u^2 \tan^2 (\pi t/2)} \right.$$
$$\left. - \frac{2 \sin \beta \pi}{\pi} \int_0^1 \frac{\tan^{2-2\beta}(\pi t/2)\, \cot^{-1}[u \tan (\pi t/2)]\, dt}{1 + u^2 \tan^2 (\pi t/2)} \right\} (1 - u^2)$$

we see that as $u \to 0$, $dx/dy \to \pm \infty$.   To examine the sign of this derivative, we substitute $u \tan (\pi t/2) = \tan (\pi s/2)$, obtaining

$$\frac{dx}{dy} = -\frac{ky}{q} + u^{2\beta-1}(1 - u^2) \left[\left(\sin \beta \pi - \frac{kh_2}{q} \cos \beta \pi \right) \int_0^1 \frac{\cot^{2\beta} (\pi s/2)\, ds}{1 + u^2 \cot^2 (\pi s/2)} \right.$$
$$\left. - \sin \beta \pi \int_0^1 \frac{s \cot^{2\beta} (\pi s/2)\, ds}{1 + u^2 \cot^2 (\pi s/2)} \right] \quad (30)$$

Then, taking the limit of this expression,

$$\lim_{u \to 0} \left(u^{1-2\beta} \frac{dx}{dy} \right) = \left(\tan \beta \pi - \frac{kh_2}{q} \right) - \sin \beta \pi\, f_1 \left(\frac{1}{2} - \beta \right)$$

we see that the sign of the derivative dx/dy is related to the sign of the quantity

$$A = \frac{kh^2}{q} - \tan \beta\pi + \sin \beta\pi \, f_1 \left(\frac{1}{2} - \beta \right)$$

that is, for $A > 0$, $dx/dy \to -\infty$; for $A < 0$, $dx/dy \to +\infty$; and for $A = 0$, dy/dy is finite. Therefore, if $A > 0$, the free surface will approach the slope BC horizontally. If $A < 0$, $dx/dy \to +\infty$, the free surface will be tangent to the downstream slope, and a surface of seepage will develop. Thus a surface of seepage can not exist if $A \geqq 0$, or

$$h_2 \geqq \left[\tan \pi\beta - f_1 \left(\frac{1}{2} - \beta \right) \sin \pi\beta \right] \frac{q}{k}$$

$$h_2 \geqq G(\beta) \frac{q}{k} \tag{31}$$

A plot of the function $G(\beta)$ as a function of β is given in Fig. 8-14.

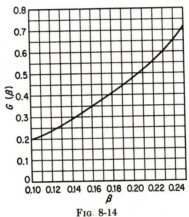

Fig. 8-14

If the inequality of (31) is not satisfied a surface of seepage will develop. In this event Numerov recommends the following approximation: First, from Eq. (31) the quantity $q/k = h_2/G(\beta)$ is determined, where h_2 is the actual tail-water height. Then, with this value for q/k in Eq. (19), the minimum value obtained for h_2 is taken as the height of the exit point of the free surface above the impervious base (h_0), as shown in Fig. 8-15.

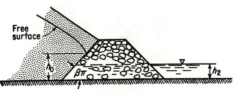

Fig. 8-15

Finally, the locus of the free surface is obtained by using the true q/k and h_0 in place of h_2 in Eq. (27b).

8-7. Seepage through Earth Dam Founded on Layer of Finite Depth with Cutoff Wall

In Fig. 8-16 we have a section of an earth dam containing an horizontal underdrain and an impervious cutoff wall. The solution to this problem

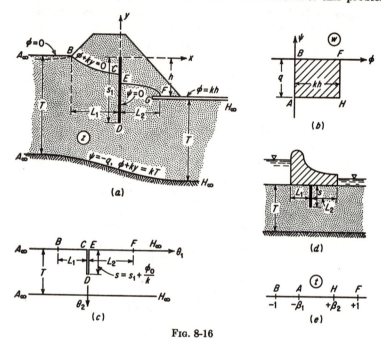

Fig. 8-16

can also be taken as an approximation for the seepage characteristics through an homogeneous dam with an impervious core (cf. Sec. 7-4). The upstream slope will be taken as horizontal and the head loss through the dam will be h. The shape of the impervious boundary is not known initially but will be determined from the condition that it is a streamline ($\psi = -q$) along which $\phi + ky = kT$, where T is the thickness of the base layer, as shown on the figure. The solution to this problem was first given by Polubarinova-Kochina, based on the solution given earlier by Voshchinin [156] for $s_1 = 0$. The same problem was also studied by Mkhitarian [95] and Nelson-Skornyakov [102]. In the latter study T was assumed to be infinite.

Defining the Zhukovsky function as

$$\theta = \theta_1 + i\theta_2 = z + \frac{iw}{k} \tag{1}$$

and designating the unknown potential at the bottom of the cutoff wall (point D) as ϕ_0, we obtain the w plane and θ plane shown in Fig. 8-16b and c, respectively. Now we note the similarity between the θ plane and Fig. 8-16d (letting $s_1 + \phi_0/k = s$, the equivalent length of sheet-pile).* Taking an auxiliary t plane, as shown in Fig. 8-16e, from the results of Sec. 5-5 we obtain for the functional relationship between the w plane and the θ plane of Fig. 8-16

$$w = kh - \frac{kh}{K} F\left(m, \sqrt{\frac{(t-1)(1+\beta_2)}{2(t-\beta_2)}}\right) \tag{2a}$$

where

$$t = \pm \cos \frac{\pi s}{2T} \sqrt{\tanh^2 \frac{\pi \theta}{2T} + \tan^2 \frac{\pi s}{2T}} \tag{2b}†$$

$$m = \sqrt{\frac{2(\beta_1 + \beta_2)}{(1+\beta_1)(1+\beta_2)}} \tag{2c}$$

$$\beta_1 = \cos \frac{\pi s}{2T} \sqrt{\tanh^2 \frac{\pi L_1}{2T} + \tan^2 \frac{\pi s}{2T}} \tag{2d}$$

$$\beta_2 = \cos \frac{\pi s}{2T} \sqrt{\tanh^2 \frac{\pi L_2}{2T} + \tan^2 \frac{\pi s}{2T}} \tag{2e}$$

Noting at point D in Fig. 8-16c that $\theta = is$, from Eq. (2b) we find that at this point $t = 0$; hence

$$\frac{\phi_0}{k} = h - \frac{h}{K} F\left(m, \sqrt{\frac{1+\beta_2}{2\beta_2}}\right) \tag{3}$$

The quantity of seepage for the dam is simply [Eq. (12), Sec. 5-5]

$$q = \frac{khK'}{K} \tag{4}$$

where the modulus is given by Eq. (2c). Values of β's can be obtained directly from Fig. 5-15. Thus, given L_1 and L_2 we first approximate the length of sheetpile s; then the *actual length* s_1 is determined from $s_1 = s - \phi_0/k$ and Eq. (3).

To obtain the equation of the free surface (along which $\psi = 0$ and $\phi + ky = 0$) we substitute into Eq. (2a) $w = -ky$ and $\theta = x$, whence we obtain

$$y = -\frac{h}{K} F\left(m, \sqrt{\frac{(1+\beta_2)(\mp|t|-1)}{2(\mp|t|-\beta_2)}}\right) \tag{5a}$$

where

$$t = \cos \frac{\pi s}{2T} \sqrt{\tanh^2 \frac{\pi x}{2T} + \tan^2 \frac{\pi s}{2T}} \tag{5b}$$

The upper sign in Eq. (5a) applies to the downstream portion of the free surface, the lower sign to the upstream part. Values of t can be obtained

* Compare also with Fig. 5-14a.
† See footnote for Eq. (5), Sec. 5-5.

directly from Fig. 5-15. To locate the intercepts of the free surface at the piling (at points C and E), with $x = 0$ in Eqs. (5) we have

$$y_C = -\frac{h}{K} F\left(m, \sqrt{\frac{(1 + \beta_2)[1 - \sin(\pi s/2T)]}{2[\beta_2 - \sin(\pi s/2T)]}}\right) \tag{6a}$$

$$y_E = -\frac{h}{K} F\left(m, \sqrt{\frac{(1 + \beta_2)[1 + \sin(\pi s/2T)]}{2[\beta_2 + \sin(\pi s/2T)]}}\right) \tag{6b}$$

Mkhitarian obtained the plots of Figs. 8-17 for two special cases. Figure 8-17a gives q/kh as a function of $\frac{s_1}{T} - \frac{h}{2T}$ for a centrally located cutoff wall ($L_1 = L_2 = L$) with $h/T = 0.25$. Figure 8-17b provides q/kh as a function of $\frac{s_1}{T} - \frac{h}{2T}$ for various ratios of h/T with $L_1/T = 0.5$ and $L_2/T = 1.0$.

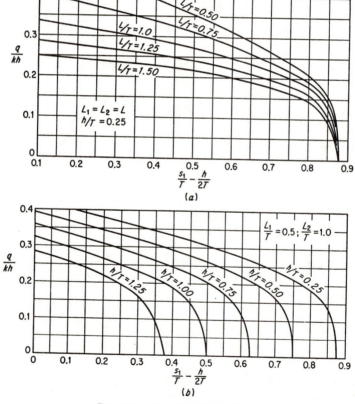

FIG. 8-17. (*After Mkhitarian* [95].)

PROBLEMS

1. Verify the mapping of Fig. 8-1c.
2. The triangular section in Fig. 8-18 is a clay core with $k = 1 \times 10^{-7}$ cm/sec. Obtain (a) the discharge, (b) the locus of the streamline through point D, and (c) the velocity distribution along the discharge slope BC.

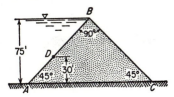

FIG. 8-18

3. In Fig. 8-4a, if $h = 50$ ft, $h_0 = 10$ ft, and $L = 50$ ft, obtain the reduced discharge through the dam.   Compare with the quantity given by Dupuit's formula.
4. Estimate the reduced discharge for each of the following:
(a) the section shown in Fig. 8-19
(b) the Cherry Valley Dam with 40 ft of tail water

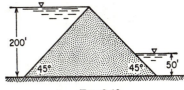

FIG. 8-19

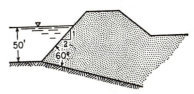

FIG. 8-20

5. Estimate the reduced discharge for the section shown in Fig. 8-20.
6. Verify Eq. (3), Sec. 8-5.
7. Demonstrate that Eq. (4), Sec. 8-6, can be reduced to Kozeny's solution (Sec. 3-2) for an earth dam with an horizontal underdrain.
8. Verify each of Eqs. (5), Sec. 8-6.
9. (a) In terms of the constants c and d [Eq. (11a), Sec. 8-6], find $W(t)$ as $t \to \infty$. (b) Verify Eq. (20), Sec. 8-6, for a dam with $\alpha = \frac{1}{2}$, $\beta = 0$, $h_1 = h$, and $h_2 = 0$.
10. For each of the dam sections shown in Fig. 8-21, obtain the reduced discharge and the locus of the free surface by (a) Numerov's method (Sec. 8-6) and (b) A. Casagrande's method (Sec. 3-2).   Compare the two.

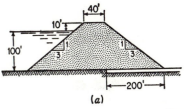

(a)   (b)

FIG. 8-21

11. Estimate the reduced discharge for the section shown in Fig. 8-22.

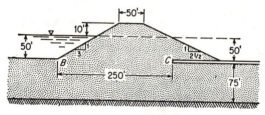

FIG. 8-22

12. Estimate the reduced discharge and obtain the locus of the free surface for the dam in Fig. 8-23.

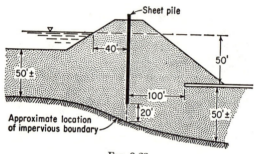

FIG. 8-23

13. Estimate the seepage per 100 ft of dam in Fig. 8-24.

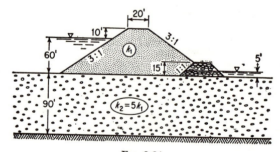

FIG. 8-24

9

Seepage from Canals and Ditches

9-1. Seepage from a Ditch with a Curved Perimeter into a Horizontal Drainage Layer

As an introduction to the subject of seepage from ditches let us investigate the characteristics of the equation

$$\theta = iz + \frac{w}{k} = Ae^{w/\alpha} \tag{1}*$$

with reference to the section shown in Fig. 9-1. Here θ is Zhukovsky's

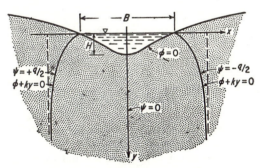

Fig. 9-1

function, $w = \phi + i\psi$, α is a parameter, and A is a real constant. Separating this expression into real and imaginary parts, we obtain

$$\begin{aligned} -y + \frac{\phi}{k} &= Ae^{\phi/\alpha} \cos \frac{\psi}{\alpha} \\ x + \frac{\psi}{k} &= Ae^{\phi/\alpha} \sin \frac{\psi}{\alpha} \end{aligned} \tag{2}$$

* This expression was given ab initio by Kozeny [76] in 1931 (see also Muskat [99]). The formal derivation of this expression was obtained by Pavlovsky [110] in 1936 by taking the θ plane of Fig. 9-1 as a circle (see Prob. 1 of this chapter).

Substituting $-\psi$ for ψ and $-x$ for x in Eqs. (2), we see that the system of streamlines defined by ψ in these equations is symmetrical about the y axis. Hence, the y axis can be taken as the streamline $\psi = 0$. Now a free surface must satisfy the conditions $-y + \phi/k = 0$ and, say, $\psi = -q/2$; from the first of Eqs. (2) we find

$$\cos\left(-\frac{q}{2\alpha}\right) = 0$$

or

$$q = -(2n + 1)\alpha\pi \tag{3}$$

where n is an integer.

In particular, taking $n = 0$ and substituting Eq. (3) with $\psi = -q/2$, and $\phi = ky$ into the second of Eqs. (2), we obtain for the free surface

$$x - \frac{q}{2k} = Ae^{-ky\pi/q} \tag{4}$$

which has the asymptote (at $y = \infty$)

$$x_{y=\infty} = \frac{q}{2k} \tag{5}$$

Letting $y = 0$ in Eq. (4), we obtain for the half width of the ditch (Fig. 9-1)

$$x_{y=0} = \frac{B}{2} = \frac{q}{2k} + A \tag{6}$$

Now, taking $\phi = 0$ in Eqs. (2), we find the parametric equations for the perimeter of the ditch:

$$-y = A \cos \frac{\psi}{\alpha}$$
$$x + \frac{\psi}{k} = A \sin \frac{\psi}{\alpha} \tag{7}$$

As $\psi = 0$ at the bottom of the ditch, we find from the first of these equations that $y = -A = H$, where H is the maximum depth of water in the ditch. Hence the quantity of seepage from the ditch section is found from Eq. (6) to be

$$q = k(B + 2H) \tag{8}$$

From Eq. (5), the width of the section at $y = \infty$ is

$$B_\infty = 2x_\infty = B + 2H \tag{9}$$

Eliminating ψ in Eqs. (7) and noting that $A = -H$, the equation of the perimeter of the ditch is

$$\pm x = -\sqrt{H^2 - y^2} + \frac{B + 2H}{\pi} \cos^{-1} \frac{y}{H} \tag{10}$$

For the velocity along the perimeter of the ditch, we have

$$V = \frac{d\phi}{ds} = \frac{1}{\sqrt{(\partial x/\partial \phi)^2 + (\partial y/\partial \phi)^2}}$$

and

$$V = \frac{k}{\sqrt{1 + [\pi H/(B + 2H)]^2 - [2\pi H/(B + 2H)] \cos [\pi \psi/k(B + 2H)]}}$$

(11)

Cross sections of the ditches for two cases are shown in Fig. 9-2. The arrows denote the velocity distribution along the perimeter of the section.

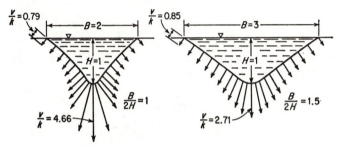

FIG. 9-2. (*After Vedernikov* [151].)

The complete flow net for $B = 2$ and $H = 1$ is given in Fig. 9-3a. From this figure we see that for practical purposes the free surface can be considered to approach its vertical asymptote, and the equipotential lines

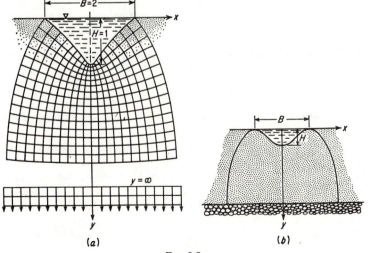

FIG. 9-3

can be taken as horizontal at a depth of $y = 3(B + 2H)/2$. Thus the solution given above may be considered as a reasonable approximation for a deep drainage layer of finite depth, as shown in Fig. 9-3b.

9-2. Seepage from a Ditch into a Curved Drainage Layer*

If we take $n = -1$ in Eq. (3), Sec. 9-1, then $\alpha = q/\pi$, and Eq. (1), Sec. 9-1, will be (changing A to $-C$)

$$-iz = \frac{w}{k} + Ce^{\pi w/q} \tag{1}$$

Thus

$$-x = Ce^{\pi\phi/q} \sin \frac{\pi\psi}{q} + \frac{\psi}{k}$$
$$y = Ce^{\pi\phi/q} \cos \frac{\pi\psi}{q} + \frac{\phi}{k} \tag{2}$$

As in the previous section we find that $C = H$, where H is the maximum depth of water in the ditch; therefore, for $\phi = 0$, we get

$$-x = H \sin \frac{\pi\psi}{q} + \frac{\psi}{k}$$
$$y = H \cos \frac{\pi\psi}{q}$$

and the perimeter of the ditch is given by

$$\pm x = \sqrt{H^2 - y^2} + \frac{q}{k\pi} \cos^{-1} \frac{y}{H} \tag{3}$$

At $y = 0$, $x = B/2$, and

$$q = k(B - 2H) \tag{4}$$

Hence

$$\pm x = \sqrt{H^2 - y^2} + \frac{B - 2H}{\pi} \cos^{-1} \frac{y}{H} \tag{5}$$

For the equation of the free surface, where $\psi = -q/2$ and $y = \phi/k$, we obtain

$$x = He^{\pi k y/q} + \frac{q}{2k} \tag{6}$$

Evidently, as $y \to +\infty$, $x \to +\infty$, and the free surface will develop for this case as shown in Fig. 9-4a.

The shapes of the ditches and their free surfaces for both cases (curve I for $n = 0$ and curve II for $n = -1$) with $B = 8$ and $H = 1$ are shown in Fig. 9-4b. In reference to this plot we note that the differences in the developed free surfaces for the two conditions depicted cannot be attributed to the small differences in the shape of the ditches but result from the nature of the boundary equipotential lines imposed at infinity. For the first case ($n = 0$) we found in Sec. 9-1 that the equipotential lines

* See previous footnote.

approach the horizontal. In the present case ($n = -1$) the boundary equipotential lines will approach circular arcs that account for the lateral spreading of the free surface. The present solution can be considered as a valid approximation for the seepage condition depicted in Fig. 9-4c.

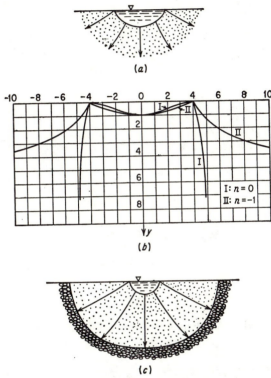

Fig. 9-4. (b) (*After Muskat* [99].)

Finally, although the solutions of this and the previous section are valid only for rather special ditch sections and idealized boundaries, it is noteworthy that in both cases only a knowledge of the width and depth of the section is required in order to provide a measure of the seepage quantity.

9-3. Seepage from Ditches of Trapezoidal Shape

A much more direct method of solution for the seepage from ditches, canals, etc., was given by Vedernikov [151], using the method of inversion. The section to be investigated is shown in Fig. 9-5a. The hodograph is given in Fig. 9-5b, and the inversion of the hodograph [$dz/dw = 1/(u - iv)$] in Fig. 9-5c.

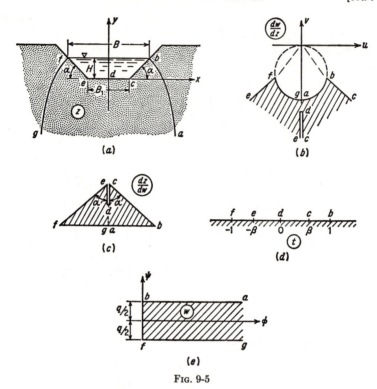

Fig. 9-5

Taking an auxiliary t plane as shown in Fig. 9-5d, we obtain for the mapping of the dz/dw plane onto the lower half plane of t

$$\frac{dz}{dw} = M \int_0^t \frac{t\,dt}{(1-t^2)^{1/2+\sigma}(\beta^2-t^2)^{1-\sigma}} + N = M\Phi(t) + N \qquad (1)$$

where $\sigma = \alpha/\pi$, and $\Phi(t)$ is the indicated integral. In particular, we shall define

$$\begin{aligned}
J_1 &= \Phi(\beta) = \int_0^\beta \frac{t\,dt}{(1-t^2)^{1/2+\sigma}(\beta^2-t^2)^{1-\sigma}} \\
J_2 &= \int_\beta^1 \frac{t\,dt}{(1-t^2)^{1/2+\sigma}(t^2-\beta^2)^{1-\sigma}}
\end{aligned} \qquad (2)$$

Substituting $(t^2-\beta^2)/(1-\beta^2) = x$ into the second of Eqs. (2), we find that (cf. Sec. B-7)

$$J_2 = \frac{1}{2\beta'} \int_0^1 x^{\sigma-1}(1-x)^{-1/2-\sigma}\,dx = \frac{1}{2\beta'}\,B(\sigma, \tfrac{1}{2}-\sigma) = \frac{\Gamma(\sigma)\Gamma(\tfrac{1}{2}-\sigma)}{2\beta'\sqrt{\pi}} \qquad (3)$$

where $\beta' = (1-\beta^2)^{1/2}$.

To eliminate the constant N in Eq. (1), we note that at points c, $t = \beta$, and the velocity is infinite ($dz/dw = 0$); hence

$$MJ_1 + N = 0$$

and
$$\frac{dz}{dw} = M[\Phi(t) - J_1] \tag{4}$$

To evaluate the constant M in Eq. (4) we note that at points b, where $u = k \sin \pi\sigma \cos \pi\sigma$, $v = -k \cos^2 \pi\sigma$, $dw/dz = u - iv = kie^{-\pi i\sigma} \cos \pi\sigma$, and $t = 1$,

$$\frac{1}{ki} e^{\pi i\sigma} = M[\Phi(1) - J_1] \cos \pi\sigma$$

Now, noting in the second of Eqs. (2) that

$$(\beta^2 - t^2)^{1-\sigma} = -e^{-i\pi\sigma}(t^2 - \beta^2)^{1-\sigma}$$

we have

$$\Phi(t) = J_1 - e^{i\pi\sigma} \int_\beta^t \frac{t \, dt}{(1 - t^2)^{\frac{1}{2}+\sigma}(t^2 - \beta^2)^{1-\sigma}}$$
$$\Phi(1) = J_1 - e^{\pi i\sigma} J_2 \tag{5}$$

hence

$$M = \frac{i}{kJ_2 \cos \pi\sigma} \tag{6}$$

and
$$\frac{dz}{dw} = \frac{i}{kJ_2 \cos \pi\sigma} [\Phi(t) - J_1] \tag{7}$$

The mapping of the w plane (Fig. 9-5e) onto the lower half of the t plane is given by (cf. Sec. 4-7)

$$w = \frac{iq}{\pi} \sin^{-1} t \tag{8}$$

Now multiplying Eq. (7) by the derivative of Eq. (8) with respect to t, we find that

$$\frac{dz}{dt} = \frac{dz}{dw} \frac{dw}{dt} = -\frac{q[\Phi(t) - J_1]}{k\pi J_2 \cos \pi\sigma \sqrt{1 - t^2}}$$

which, after integration with respect to t, yields

$$z = -\frac{q}{k\pi J_2 \cos \pi\sigma} \left[\int_0^t \frac{\Phi(t) \, dt}{\sqrt{1 - t^2}} - J_1 \sin^{-1} t \right] \tag{9a}$$

For the integral in Eq. (9a) we integrate by parts.

$$\int_0^t \frac{\Phi(t) \, dt}{\sqrt{1 - t^2}} = \Phi(t) \sin^{-1} t - \int_0^t \frac{t \sin^{-1} t \, dt}{(1 - t^2)^{\frac{1}{2}+\sigma}(\beta^2 - t^2)^{1-\sigma}} \tag{9b}$$

We shall now consider Eqs. (9) for the various parts of the flow region.

Along the bottom of the ditch, where $0 < t < \beta$,

$$z = -\frac{q}{k\pi J_2 \cos \pi\sigma} \left\{ \sin^{-1} t \left[\int_0^t \frac{t \, dt}{(1-t^2)^{\frac{1}{2}+\sigma}(\beta^2-t^2)^{1-\sigma}} - J_1 \right] \right.$$
$$\left. - \int_0^t \frac{t \sin^{-1} t \, dt}{(1-t^2)^{\frac{1}{2}+\sigma}(\beta^2-t^2)^{1-\sigma}} \right\} \quad (10a)$$

At points c, where $t = \beta$ and $z = B_1/2$, we find

$$\frac{B_1}{2} = \frac{q}{\pi k J_2 \cos \pi\sigma} \int_0^\beta \frac{t \sin^{-1} t \, dt}{(1-t^2)^{\frac{1}{2}+\sigma}(\beta^2-t^2)^{1-\sigma}} \quad (10b)$$

Along the side of the ditch bc, where $\beta < t < 1$,

$$z = \frac{B_1}{2} + \frac{q}{k\pi J_2 \cos \pi\sigma} e^{\pi\sigma i} \left[\sin^{-1} t \int_\beta^t \frac{t \, dt}{(1-t^2)^{\frac{1}{2}+\sigma}(t^2-\beta^2)^{1-\sigma}} \right.$$
$$\left. - \int_\beta^t \frac{t \sin^{-1} t \, dt}{(1-t^2)^{\frac{1}{2}+\sigma}(t^2-\beta^2)^{1-\sigma}} \right] \quad (11a)$$

At points b, where $t = 1$ and $z = B/2 + iH$, we obtain

$$\frac{B-B_1}{2} = \frac{q}{k\pi J_2} \left[\frac{\pi}{2} J_2 - \int_\beta^1 \frac{t \sin^{-1} t \, dt}{(1-t^2)^{\frac{1}{2}+\sigma}(t^2-\beta^2)^{1-\sigma}} \right] \quad (11b)$$

$$H = \frac{q}{k\pi J_2} \tan \pi\sigma \left[\frac{\pi}{2} J_2 - \int_\beta^1 \frac{t \sin^{-1} t \, dt}{(1-t^2)^{\frac{1}{2}+\sigma}(t^2-\beta^2)^{1-\sigma}} \right] \quad (11c)$$

Along the free surface ba, where $1 < t < \infty$, from Eq. (11a) we find

$$z = \frac{B}{2} + Hi + \frac{q}{k\pi J_2 \cos \pi\sigma} \left[-\cosh^{-1} t \int_1^t \frac{t \, dt}{(t^2-1)^{\frac{1}{2}+\sigma}(t^2-\beta^2)^{1-\sigma}} \right.$$
$$\left. - iJ_2 e^{\pi\sigma i} \cosh^{-1} t + \int_1^t \frac{t \cosh^{-1} t \, dt}{(t^2-1)^{\frac{1}{2}+\sigma}(t^2-\beta^2)^{1-\sigma}} \right] \quad (12a)*$$

Separating this equation into real and imaginary parts, we obtain for the equation of the free surface ba,

$$x - \frac{B}{2} = \frac{q}{k\pi J_2 \cos \pi\sigma} \left[\int_1^t \frac{t \cosh^{-1} t \, dt}{(t^2-1)^{\frac{1}{2}+\sigma}(t^2-\beta^2)^{1-\sigma}} + J_2 \sin \pi\sigma \cosh^{-1} t \right.$$
$$\left. - \cosh^{-1} t \int_1^t \frac{t \, dt}{(t^2-1)^{\frac{1}{2}+\sigma}(t^2-\beta^2)^{1-\sigma}} \right] \quad (12b)$$

We shall now derive the expression for the discharge from the ditch. Defining

$$\int_0^\beta \frac{t \sin^{-1} t \, dt}{(1-t^2)^{\frac{1}{2}+\sigma}(\beta^2-t^2)^{1-\sigma}} = f_1(\sigma,\beta)$$
$$\int_\beta^1 \frac{t \sin^{-1} t \, dt}{(1-t^2)^{\frac{1}{2}+\sigma}(\beta^2-t^2)^{1-\sigma}} = f_2(\sigma,\beta) \quad (13)$$

---

* We note that $(1-t^2)^{\frac{1}{2}+\sigma} = ie^{\pi\sigma i}(t^2-1)^{\frac{1}{2}+\sigma}$. Also for real values of $t > 1$, $\sin^{-1} t = \pi/2 - i \cosh^{-1} t$.

we have, in place of Eq. (10b),

$$B_1 = \frac{2q}{k\pi J_2 \cos \pi\sigma} \, f_1(\sigma,\beta) \tag{14a}$$

and in place of Eqs. (11b) and (11c),

$$\frac{B - B_1}{2} = \left[\frac{\pi}{2} J_2 - f_2(\sigma,\beta)\right] \frac{q}{\pi k J_2}$$
$$H = \frac{q}{\pi k J_2} \tan \pi\sigma \left[\frac{\pi}{2} J_2 - f_2(\sigma,\beta)\right] \tag{14b}$$

whence

$$B = B_1 + \frac{q}{k}\left[1 - \frac{2f_2(\sigma,\beta)}{\pi J_2}\right]$$
$$H = \frac{q}{2k} \tan \sigma\pi \left[1 - \frac{2f_2(\sigma,\beta)}{\pi J_2}\right] \tag{14c}$$

We note in Eqs. (14c) that the quantity of seepage is dependent upon the parameters σ and β and one of the dimensions B, B_1, or H, which are related by $B - B_1 = 2H \cot \sigma\pi$. As was done in the previous sections, Vedernikov takes the quantity of seepage in the form

$$q = k(B + AH) \tag{15a}$$

where, from Eqs. (14c), A is given by

$$A = \frac{2}{\tan \sigma\pi} \frac{f_2(\sigma,\beta) - f_1(\sigma,\beta)/\cos \sigma\pi}{J_2\pi/2 - f_2(\sigma,\beta)} \tag{15b}$$

Taking a series of values for α and β, Vedernikov obtained the correspondence between A and B/H as given in Fig. 9-6. In this figure $m = \cot \alpha$ is the side slope of the ditch.

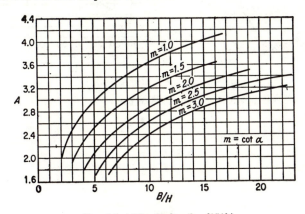

FIG. 9-6. (*After Vedernikov* [151].)

Comparing Eq. (15a) with the expression for the quantity of seepage obtained in Sec. 9-1 for ditches with a curved perimeter $[q = k(B + 2H)]$,

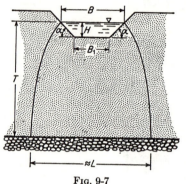

we see that the latter solution implied $A = 2$, whereas in Fig. 9-6, for the trapezoidal section, A is seen to vary from 2 to 4 for typical values of B/H. Noting that the velocity at infinity equals the coefficient of permeability, we find that the width of the flow at infinity (Fig. 9-7) is

$$L = B + AH \qquad (16)$$

Thus we see that for a trapezoidal section, as was also shown to be the case for the curved perimeter, the equipotential lines rapidly approach

Fig. 9-7

the horizontal. Hence the solution of this section may also be considered to provide a sufficiently valid approximation for seepage into deep horizontal filters, as shown in Fig. 9-7.

9-4. Seepage from Triangular-shaped Ditches

In the case of a triangular ditch (Fig. 9-8a), the solution of the previous

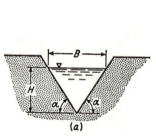

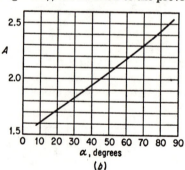

Fig. 9-8. (b) (*After Vedernikov* [151].)

section is modified by taking $\beta = 0$. We then have

$$q = k(B + AH) \qquad (1)$$

where [Eq. (15b), Sec. 9-3]

$$A = \frac{2}{\tan \sigma\pi} \frac{f_2(\sigma)}{J_2\pi/2 - f_2(\sigma)} \qquad (2)$$

$$\sigma = \frac{\alpha}{\pi}$$

$$J_2 = \frac{\Gamma(\sigma)\Gamma(\frac{1}{2} - \sigma)}{2\sqrt{\pi}}$$

A plot of the relationship between A and α was obtained by Vedernikov and is given in Fig. 9-8b.

9-5. Seepage from Ditches into Permeable Layers at Shallow Depths

In this section we shall investigate the influence of the depth of horizontal permeable layers upon the quantity of seepage from ditches. The

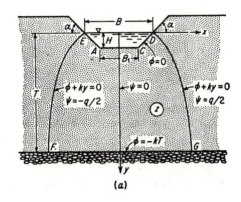

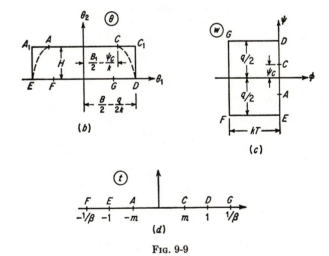

Fig. 9-9

solution given here is based upon that first given by Vedernikov [151] in 1934.*

The section to be investigated is shown in Fig. 9-9a. Again taking the Zhukovsky function as $\theta = z + iw/k$, we have, as the corresponding

---

* The solution has also been given by Muskat [99].

θ plane, the solid rectangle in Fig. 9-9b. In actuality, the images of the sides of the ditch (CD and AE) are not known initially in the θ plane (shown dotted; for example at point C, $\theta_1 = B_1/2 - \psi_c/k$). The assumption of a rectangular θ plane greatly reduces the mapping problem and appears justifiable as the results obtained from the solution given here differ but little from the few available data where more sophisticated analyses were employed [138, 154]. The w plane is shown in Fig. 9-9c.

Selecting the t plane of Fig. 9-9d, for the mapping of the θ plane, we find

$$\theta = M \int_0^{t/m} \frac{dt}{\sqrt{(1 - t^2)(1 - m^2 t^2)}} + iH \tag{1a}$$

From the correspondence at points D where $t = 1$ and $\theta = B/2 - q/2k$, we obtain

$$\frac{B - q/k}{2} - iH = M(K + iK') \tag{1b}$$

and hence

$$q = k\left(B + \frac{2HK}{K'}\right) \tag{2}$$

The mapping of the w plane onto the t plane is given by

$$w = M_1 \int_0^t \frac{dt}{\sqrt{(1 - t^2)(1 - \beta^2 t^2)}} \tag{3a}$$

where the constant M_1 is found (points D) to be

$$M_1 = \frac{iq}{2K(\beta)} \tag{3b}$$

Hence from the correspondence at points G, where $w = -kT + iq/2$ and $t = 1/\beta$, we obtain

$$q = \frac{2kTK(\beta)}{K'(\beta)} \tag{4}$$

As noted from Eqs. (2) and (4), one more expression relating the unknowns q/k, m, and β is required. At points C, where

$$\theta = \frac{B_1}{2} - \frac{\psi_c}{k} + iH$$

$t = m$, and $w = i\psi_c$, substituting into Eq. (1a), we get

$$\psi_c = \frac{k(q/k - B + B_1)}{2} \tag{5}$$

From Eq. (3a)

$$\psi_c = \frac{q}{2K(\beta)} F(\beta, \phi) \tag{6}$$

where $F(\beta,\phi)$ is the elliptic integral of the first kind of modulus β and amplitude $\phi = \sin^{-1} m$. Equating Eqs. (5) and (6) and solving for q/k, we find finally that

$$\frac{q}{k} = \frac{2Hn}{1 - F(\beta,\phi)/K(\beta)} \tag{7}$$

where $n = \cot \alpha = (B - B_1)/2H$.

Thus all the required information necessary to effect a solution is available in Eqs. (2), (4), and (7).* A plot of q/kH as a function of B/H

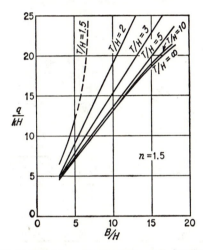

FIG. 9-10. (*Adapted from Muskat* [99] *and Vedernikov* [151].)

for various ratios of T/H, with $n = 1.5$, is shown in Fig. 9-10. (Vedernikov gave B/H versus $2K/K'$; Muskat applied Eq. (2) to obtain q/kH versus B/H.)

9-6. Seepage from a Shallow Ditch Considering Capillarity†

The section to be considered is shown in Fig. 9-11a. Here MM' $(= B)$ represents the actual width of the canal, and LL' $(= B_1)$ gives the effective width of the section assuming a *height of capillary rise* (cf. Sec. 1-7) of h_c. Also, it will be assumed that the water level in the ditch is small in comparison to its width; that is, $B/H \gg 1$.

The velocity hodograph of the right half of the flow region is given

* Assuming q/kT, β is obtained from Eq. (4). With β known, m can be found from Eq. (7) and then B/H from Eq. (2).

† This solution was first derived by Risenkampf [122] in 1940.

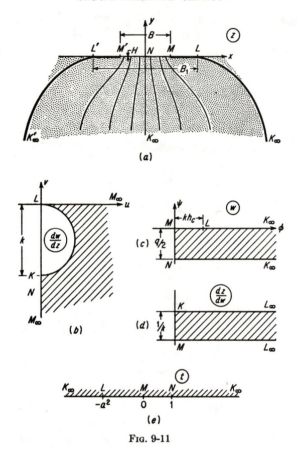

Fig. 9-11

in Fig. 9-11b, and the w plane in Fig. 9-11c. The inverse of the hodograph plane will be a semi-infinite strip (Fig. 9-11d).

For the mapping of the w plane onto the t plane (Fig. 9-11e), we find

$$t = \sin^2 \frac{\pi i w}{q} = -\sinh^2 \frac{\pi w}{q} \tag{1}$$

whence, from the correspondence at points $L(t = -a^2, w = kh_c)$, we get

$$a = \sinh \frac{\pi k h_c}{q} \tag{2}$$

Mapping the inverse of the hodograph plane onto the t plane,

$$\frac{1}{u - iv} = \frac{dz}{dw} = M \int_0^t \frac{dt}{(t + a^2)t^{1/2}} + N = \frac{2M}{a} \tan^{-1} \frac{\sqrt{t}}{a} + N \tag{3}$$

and noting that at points M, $t = 0$ and $u - iv = \infty$, and at points K,

$t = \infty$ and $v = -k$, we obtain

$$\frac{dz}{dw} = -\frac{2i}{k\pi} \tan^{-1} \frac{\sqrt{t}}{a} = -\frac{2i}{k\pi} \tan^{-1} \left[\frac{\sin (\pi i w/q)}{a} \right] \qquad (4)$$

Defining $\pi i w/q = W$, Eq. (4) can be represented by the series

$$\frac{dz}{dw} = -\frac{4i}{k\pi} \sum_{n=1}^{\infty} e^{-(2n-1)\alpha} \frac{\sin (2n-1)W}{2n-1} \qquad (5)*$$

where $\alpha = \sinh^{-1} a = \pi k h_c/q$. Integrating Eq. (5) and noting that $z = 0$ for $w = -iq/2$, we get

$$z = \frac{4q}{k\pi^2} \sum_{n=1}^{\infty} e^{(-2n-1)\alpha} \frac{\cos (2n-1)W}{(2n-1)^2} \qquad (6)$$

Now for points M, where $z = B/2$ and $w = W = 0$, we find that the width of the ditch is

$$B = \frac{8q}{k\pi^2} \sum_{n=1}^{\infty} \frac{e^{-(2n-1)\alpha}}{(2n-1)^2} \qquad (7)$$

and for points L, where $z = B_1/2$, $w = k h_c$, and $W = i\alpha$, the effective width of the ditch is

$$B_1 = \frac{8q}{k\pi^2} \sum_{n=1}^{\infty} e^{-(2n-1)\alpha} \frac{\cosh (2n-1)\alpha}{(2n-1)^2} \qquad (8a)$$

Or, expressing the hyperbolic cosine in exponential form,

$$B_1 = \frac{4q}{k\pi^2} \sum_{n=1}^{\infty} \frac{1 + e^{-2(2n-1)\alpha}}{(2n-1)^2} \qquad (8b)$$

Noting that at $y = -\infty$ (at point K) the velocity is equal to the coefficient of permeability, the maximum width of the flow region is simply

$$B_\infty = \frac{q}{k} \qquad (9)$$

Also we note that in the absence of capillarity ($h_c = 0$), $\alpha = 0$, and that the discharge q_0 for this case is simply

$$q_0 = kB \qquad (10)\dagger$$

---

* This series was obtained from

$$\tan^{-1} \left[\frac{2p(\sin x)}{(1-p^2)} \right] = 2 \sum_{n=1}^{\infty} p^{2n-1} \frac{[\sin (2n-1)x]}{(2n-1)}$$

† This equation is derived from Eq. (7), noting $\sum_{n=1}^{\infty} 1/(2n-1)^2 = \pi^2/8$.

Curves for q/q_0, B_1/B, and B_∞/B as a function of h_c/B are shown in Fig. 9-12. To illustrate the use of these curves, suppose we have a ditch 10 ft wide, and the height of capillary rise is estimated as 1 ft. Entering Fig. 9-12 with $h_c/B = 0.1$, we see that the discharge will be increased by

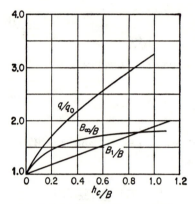

FIG. 9-12. (*After Risenkampf* [122].)

approximately 40 per cent, the effective width of the ditch will be approximately 10.9 ft ($B_1/B = 1.09$), and the width of the flow domain at great depths will be increased approximately 30 per cent.

9-7. Seepage from a Ditch into a Permeable Layer of Finite Length*

The section to be investigated is shown in Fig. 9-13a. Here again we shall assume that $B/H \gg 1$. Also, it will be assumed that L_{dr} can be approximated with sufficient accuracy from Fig. 7-12b so that the length L can be considered as a known quantity. With the Zhukovsky function as $\theta = z + iw/k$, the θ plane is as shown in Fig. 9-13b. The w plane is represented in Fig. 9-13c.

The mapping of the w plane onto the θ plane is given by

$$w = M \int_0^\theta \frac{d\theta}{\sqrt{\theta(\theta - m^2)(\theta - 1)}} - kT \tag{1}$$

Substituting $\theta = m^2 t^2$ (the t plane is given in Fig. 9-13d), we obtain

$$w = 2M \int_0^t \frac{dt}{\sqrt{(1 - t^2)(1 - m^2 t^2)}} - kT \tag{2}$$

Considering the correspondence at points D, where $t = 1$ and

$$w = -kT + iq$$

* This section is based on a solution given by Vedernikov [152].

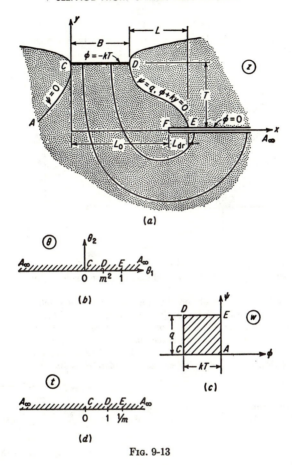

Fig. 9-13

we get $M = iq/2K$; and then from points E, where $t = 1/m$ and $w = iq$, we obtain

$$q = \frac{kTK}{K'} \tag{3}$$

Finally, noting that $\theta = z + iw/k = m^2 t^2$, from the correspondence between the θ plane and the t plane at points D and E we find

$$\frac{B}{T} = \frac{m^2}{m'^2} \frac{L}{T} + \frac{K}{K'} \tag{4}$$

from which the modulus m can be determined for a given B/T and L/T. For example, assuming a series of values for m with L/T known, a plot of B/T versus m and hence the value of the modulus corresponding to the given B/T can be obtained.

PROBLEMS

1. Taking the Zhukovsky function $\theta = iz + w/k$ to be a semicircle of radius H (Fig. 9-14), obtain Eq. (1) of Sec. 9-1.

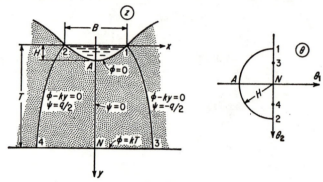

FIG. 9-14

2. If $T \to \infty$, determine the velocity at $y = \infty$ in Fig. 9-1.

3. Examine the error associated with the assumption made in the last paragraph of Sec. 9-1.

4. On the basis of Sec. 9-1 obtain the cross section of the ditch and the velocity distribution along its perimeter for (a) $B = \pi - 2$, $H = 1$ and (b) $B = 20$, $H = 2$.

5. Demonstrate that the boundary equipotential lines in Sec. 9-2 approach circular arcs as they move away from the ditch section.

6. Solve Prob. 4 for a trapezoidal ditch by the method of Sec. 9-3 with $\cot \alpha = 1$.

7. Obtain A [Eq. (15b), Sec. 9-3] if $\cot \alpha = 4$ and $B/H = 10$.

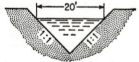

FIG. 9-15

8. Obtain the reduced quantity of seepage and the velocity distribution along the sides of the ditch in Fig. 9-15.

9. In Fig. 9-9a, $B = 25$ ft, $H = 5$ ft, and $B_1 = 15$ ft. Obtain (a) a plot of q/kH as a function of T/H and (b) the locus of the free surface (DG) if $T = 25$ ft.

10. Noting that $\sum_{n=1}^{\infty} 1/(2n - 1)^2 = \pi^2/8$, show that Eqs. (7) and (8), Sec. 9-6,

express the functional relationship $B_1(\alpha) = q/2k + B(2\alpha)/2$.

11. If the height of capillary rise in Prob. 6 is $h_c = 1$ ft, estimate the reduced quantity of seepage.

12. In Fig. 9-13 if $B = 20$ ft, $L_0 = 40$ ft, and $T = 25$ ft, obtain (a) the reduced quantity of seepage and (b) the locus of the free surface DE.

13. If an equipotential line in Fig. 9-13a is taken as the bottom of a ditch, the analysis of Sec. 9-7 may be extended to include other than shallow sections. Find the relationship between H (depth of the ditch, measured vertically at $x = B/2$) and the characteristic dimensions of the section.

14. Obtain the solution for Fig. 9-13a assuming the right half of the section to be symmetrical about the y axis.

10

Seepage toward Wells

10-1. Introduction: Fundamental Equations; Sources and Sinks

To the English-speaking engineer, the subject of seepage toward wells represents a singular phase of the seepage problem. Unlike the studies of flow through earth dams, seepage under hydraulic structures, etc., there exists in the English language a systematic development of the solutions to well problems (cf. Hall [46]). The works of Muskat [99] and Luthen et al. [88] far and away represent the most complete treatment of the subject available in any language.* The latter also provides an excellent bibliography of recent developments.

In the work of this chapter we shall present some of the fundamental aspects of well problems. The field of seepage toward wells is so vast that we can at best consider only a few representative problems which tend to illustrate the methods of solution. We shall begin by deriving the equations for radial flow.

The fundamental equation governing two-dimensional steady flow in rectangular coordinates,

$$\nabla^2\phi = \frac{\partial^2\phi}{\partial x^2} + \frac{\partial^2\phi}{\partial y^2} = 0$$

becomes, in cylindrical coordinates,

$$\nabla^2\phi = \frac{1}{r}\frac{\partial}{\partial r}\left(r\frac{\partial\phi}{\partial r}\right) + \frac{1}{r^2}\frac{\partial^2\phi}{\partial\theta^2} = 0 \tag{1}$$

where $r = (x^2 + y^2)^{1/2}$ and $\theta = \tan^{-1}(y/x)$. For pure radial flow in the horizontal plane ($z = x + iy = r\exp i\theta$) there is no variation with θ and Eq. (1) reduces to

$$\frac{1}{r}\frac{\partial}{\partial r}\left(r\frac{\partial\phi}{\partial r}\right) = 0$$

so that

$$\phi = C_1\ln r + C_2 \tag{2}$$

---

* See Charny [18] in Russian.

Now, characterizing a radial flow system through a homogeneous layer of thickness T by

$$r = r_w \qquad \text{when } \phi = 0$$
$$r = R \qquad \text{when } \phi = kh$$

where r_w = radius of well

R = radius of influence (see Sec. 2-8)

h = head loss

we find for Eq. (2)

$$\phi = \frac{kh}{\ln (R/r_w)} \ln \frac{r}{r_w} \tag{3}$$

and for the radial velocity

$$v_r = \frac{\partial \phi}{\partial r} = \frac{kh}{r \ln (R/r_w)}$$

Evidently the quantity of seepage at any distance r from the center of the well will be

$$Q = T \int_0^{2\pi} r v_r \, d\theta = \frac{2\pi T k h}{\ln (R/r_w)} \tag{4}$$

where T is the thickness of the permeable layer. We note that if

$$h = h_2 - h_1$$

and $T = (h_2 + h_1)/2$, Eq. (4) yields Eq. (3), Sec. 2-8. Substituting Eq. (3) above into Eq. (4), we obtain

$$\phi = \frac{Q}{2\pi T} \ln \frac{r}{r_w} \tag{5a}$$

Associated with the potential function of Eq. (5a) is the stream function

$$\psi = \frac{Q}{2\pi T} \theta \tag{5b}*$$

Expressing z as $re^{i\theta}$, Eqs. (5) can be combined to yield the complex potential

$$w = \phi + i\psi = \frac{Q}{2\pi T} \ln z + C \tag{6}$$

where C is a constant.

The radial pattern of flow associated with Eq. (6) is said to be either a *source* (for $+Q$) or a *sink* (for $-Q$). The term $Q/2\pi T$ with $T = 1$ is called the *strength of the source (or sink)*. Although sources and sinks are simply mathematical conveniences, they are of great value in that, when combined with other simple patterns, they can reproduce closely

---

* This equation is derived from $v_r = \partial \phi / \partial r = (1/r)(\partial \psi / \partial \theta)$.

many complicated natural flow patterns. The power of the method of sources and sinks stems from the linearity of Laplace's equation. For, if we add together a number of complex potentials such as Eq. (6), each of which satisfies $\nabla^2 w = 0$, the sum of these will satisfy Laplace's equation.

10-2. Well and Uniform Flow

As an example of the method of sources and sinks, we shall consider a uniform flow of velocity U past a well in a layer of thickness T. For convenience the well will be located at the origin (Fig. 10-1). Noting

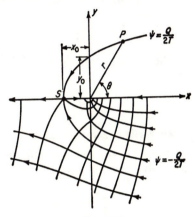

Fig. 10-1

that the complex potential for the well is $w = Q \ln z / 2\pi T + C$, and the complex potential for uniform flow is Uz, we obtain for the complex potential for the combined flow

$$w = \frac{Q \ln z}{2\pi T} + Uz + C \tag{1}$$

and hence, the potential and stream functions are

$$\phi = Ux + \frac{Q}{2\pi T} \ln r + C \qquad \psi = Uy + \frac{Q}{2\pi T} \theta$$

or $\qquad \phi = Ux + \frac{Q}{4\pi T} \ln \frac{x^2 + y^2}{r_w^2} \qquad \psi = Uy + \frac{Q}{2\pi T} \tan^{-1} \frac{y}{x} \tag{2}$

To obtain the x coordinate x_0 of the point of stagnation S, we set the velocity in the x direction equal to zero,

$$u = \frac{\partial \phi}{\partial x} = U + \frac{Q}{2\pi T} \frac{x}{x^2 + y^2} = 0$$

whence, with $y = 0$, we obtain

$$x_0 = -\frac{Q}{2\pi TU} \tag{3}$$

The streamline passing through any point P (Fig. 10-1) is found from the second of Eqs. (2) to be

$$\psi = Ur \sin \theta + \frac{Q}{2\pi T} \theta$$

so that, at the stagnation point, where $\theta = \pm\pi$, the streamlines are

$$\psi = \pm \frac{Q}{2T} \tag{4a}$$

and the y ordinates of the limiting streamlines defining the water divide are

$$y_0 = \pm \frac{Q}{4TU} \tag{4b}$$

10-3. Flow between Two Wells of Equal Strength

Consider a source and a sink of equal strengths ($Q/2\pi T = q/2\pi$) spaced at $-d$ and $+d$, respectively, from the origin on the x axis (Fig. 10-2).

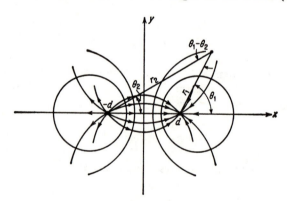

Fig. 10-2

The resulting complex potential will be

$$w = w_{\text{source}} + w_{\text{sink}}$$
$$w = -\frac{q}{2\pi} \ln (z + d) + \frac{q}{2\pi} \ln (z - d) = \frac{q}{2\pi} \ln \frac{z - d}{z + d} \tag{1}$$

whence, taking $z - d = r_1 e^{i\theta_1}$ and $z + d = r_2 e^{i\theta_2}$, we obtain for the potential and stream functions

$$\phi = \frac{q}{2\pi} \ln \frac{r_1}{r_2} \qquad \psi = \frac{q}{2\pi} (\theta_1 - \theta_2) \tag{2}$$

To find the equation of the equipotential lines we note that if $\phi = $ constant, $r_1/r_2 = $ constant. Hence, setting

$$\left(\frac{r_1}{r_2}\right)^2 = \frac{(x - d)^2 + y^2}{(x + d)^2 + y^2} = \text{constant} = C \tag{3}$$

after some elaboration, the equation of the equipotential lines is found to be

$$\left(x - d\frac{1 + C}{1 - C}\right)^2 + y^2 = \frac{4d^2C}{(1 - C)^2} \tag{4}$$

Equation (4) demonstrates that the equipotential lines are a family of circles with radii $2d\sqrt{C}/(1 - C)$, and with centers at

$$x = d\frac{1 + C}{1 - C} \qquad y = 0$$

To obtain the equation of the streamlines, we set

$$\tan \frac{2\pi\psi}{q} = \tan(\theta_1 - \theta_2) = \text{constant} = C'$$

whence, noting that $\theta_1 = \tan^{-1}[y/(x - d)]$ and $\theta_2 = \tan^{-1}[y/(x + d)]$, we find

$$x^2 + \left(y - \frac{d}{C'}\right)^2 = d^2\left(1 + \frac{1}{C'^2}\right) \tag{5}$$

Equation (5) indicates that the streamlines are also a family of circles with radii $d[1 + (1/C'^2)]^{1/2}$ and with centers at $x = 0$, $y = d/C'$.

The physical flow pattern represented by the combination of a source and a sink closely resembles the pattern of flow between a pumped well and a recharge well, for example, where water is pumped continuously from one well, used as an industrial coolant, and then is returned to the ground through a recharge well. The same flow pattern was used by Muskat [99] to simulate flow from a line source, represented by the y axis in Fig. 10-2, into a well at $x = d$.

10-4. An Eccentrically Placed Well within a Circular Contour*

In Fig. 10-3, M_1 represents the center of a (source) well of small radius r_w with a discharge of $q_1 = Q_1/T$, located at a distance e from the center

* This analysis and that of the next section are based on that given by Polubarinova-Kochina [116]. The solution of this section can also be obtained directly from the results of Eq. (3), Sec. 10-3 (cf. Muskat [99]).

of the circle of influence of radius R. The coordinates of the well are $Z_1 = X_1 + iY_1$. Point M_2 is the inverse (Sec. 3-6) of point M_1 about the circle of influence, $Z_2 = R^2/Z_1$. A (sink) well with the same intensity

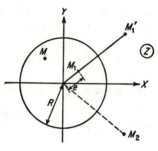

Fig. 10-3

as M_1 is to be located at point M_1' (the conjugate of point M_2) with the complex coordinates $R^2/\bar{Z}_1$. Point M is an arbitrary point within the circular contour having a complex potential

$$w(Z) + \phi(X,Y) + i\psi(X,Y) \qquad (1)$$

From Eq. (1), Sec. 10-3, we find the potential at point M, considering the source at M_1 and the sink at M_1', is

$$w(Z) = \frac{q_1}{2\pi} \ln \frac{Z - Z_1}{Z - (R^2/\bar{Z}_1)} + C \qquad (2)$$

where $\bar{Z}_1 = X_1 - iY_1$. Eq. (2) can also be written as

$$w = \frac{q_1}{2\pi} \left[\ln (Z - Z_1) - \ln \frac{R^2 - Z\bar{Z}_1}{R} \right] + C_1 \qquad (3)$$

the real part of which yields

$$\phi(X,Y) = \frac{q_1}{2\pi} \ln \frac{R|Z - Z_1|}{|R^2 - Z\bar{Z}_1|} + C_2 \qquad (4)$$

In the near vicinity of the well, at point M_1, the equipotential lines differ but little from concentric circles with M_1 as center. Assuming that such a circle can be taken as the periphery of a well with a radius r_w at M_1, from Eq. (4) we obtain for the equipotential along this circle $(Z = Z_1 + r_w)$

$$\phi_w = \frac{q_1}{2\pi} \ln \frac{Rr_w}{R^2 - |Z_1|^2} + C_2 \qquad (5)*$$

* $Z\bar{Z}_1 = (Z_1 + r_w)(\bar{Z}_1) \approx |Z_1|^2$, as the well is of small radius r_w.

For the potential function along the circle of influence ϕ_R where $Z = R$, from Eq. (4) we find

$$\phi_R = C_2 \tag{6}$$

Therefore, noting in Eq. (5) that $|Z_1|^2 = e^2$, we obtain

$$\phi_R - \phi_w = \frac{q_1}{2\pi} \ln \frac{R^2 - e^2}{Rr_w}$$

or

$$q_1 = \frac{2\pi(\phi_R - \phi_w)}{\ln (R^2 - e^2) - \ln Rr_w} \tag{7}$$

For a well with radius r_w at the center of the contour ($e = 0$) the discharge will be

$$q_0 = \frac{2\pi(\phi_R - \phi_w)}{\ln R - \ln r_w} \tag{8}$$

Dividing Eq. (7) by Eq. (8), we obtain the ratio

$$\frac{q_1}{q_0} = \frac{\ln (R/r_w)}{\ln [R/r_w - (e/r_w)^2(r_w/R)]} \tag{9}$$

which is plotted in Fig. 10-4 as a function of e/r_w for $R/r_w = 100$. It is seen from this plot that the influence of the displacement of a well

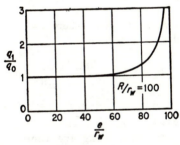

Fig. 10-4. (*After Polubarinova-Kochina* [116].)

from the center of the flow region is negligible unless the well is very close to the boundary of the contour. Thus, the assumption of simple radial flow within a circular contour appears to be justifiable even for gross displacements from the center of the contour.

10-5. Influence of the Shape of the Contour on the Discharge

In the previous section it was shown that the discharge for an eccentrically placed well with a circular contour differed but little from that of a well at the exact center of a circular boundary unless the eccentricity approached the order of the contour radius. In this section we shall extend the validity of simple radial flow to include other boundary configurations as well.

We assume that the interior of the circular contour $|Z| = R$ of Fig. 10-3 can be mapped conformally into the region bounded by the contour

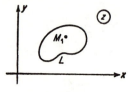

Fig. 10-5

L of Fig. 10-5 by the transformation

$$Z = F(z) \tag{1}$$

In this transformation the well of small radius r_w at M_1 of Fig. 10-3 will have as its image M_1 in Fig. 10-5 at $z = z_1$. Substituting Eq. (1) into Eq. (4), Sec. 10-4, we obtain for the potential function

$$\phi(x,y) = \frac{q_1}{2\pi} \ln \frac{R|F(z) - F(z_1)|}{|R^2 - F(z)\overline{F(z)}|} + C \tag{2}$$

Now, taking $\phi = \phi_w$ at the well, where $z = z_1 + r_w$, and considering r_w to be of small order, Eq. (2) becomes

$$\phi_w = \frac{q_1}{2\pi} \ln \frac{Rr_w|F'(z_1)|}{R^2 - |F(z_1)|^2} + C \tag{3}$$

as $F(z_1 + r_w) - F(z_1) = r_wF'(z_1) + r_w{}^2F''(z_1)/2 + \cdots \approx r_wF'(z_1)$, and $F(z_1 + r_w)\overline{F(z_1)} = |F(z_1)|^2 + r_wF'(z_1)\overline{F(z_1)} + \cdots \approx |F(z_1)|^2$.

Taking an arbitrary point along the contour of L as z_0, where $\phi = \phi_L = \phi_R$,

$$\phi_R = \frac{q_1}{2\pi} \ln \frac{R|F(z_0) - F(z_1)|}{|R^2 - F(z_0)\overline{F(z_1)}|} + C \tag{4}$$

Noting that when $|Z| = R$ in Fig. 10-3, Eq. (1) specifies $|F(z)| = R$, for a point z_0 on the contour L, $|F(z_0)| = R$ and

$$\frac{R|F(z_0) - F(z_1)|}{|R^2 - F(z_0)\overline{F(z_1)}|} = 1 \quad \text{and} \quad C = \phi_R \tag{5}$$

Consequently Eq. (3) becomes

$$\phi_R - \phi_w = \frac{q_1}{2\pi} \ln \frac{R^2 - |F(z_1)|^2}{Rr_w|F'(z_1)|}$$

and the discharge, where $F(z)$ is the function mapping the contour L onto a unit circle ($R = 1$), is

$$q_1 = \frac{2\pi(\phi_R - \phi_w)}{\ln[1 - |F(z_1)|^2] - \ln[r_w|F'(z_1)|]} \tag{6}$$

To illustrate the method of solution and also to investigate the influence of the shape of the boundary, we consider the case of a well within an elliptical contour with major and minor semiaxes of a and b, respectively. The function mapping the interior of an ellipse $x^2/a^2 + y^2/b^2 = 1$ onto the interior of a unit circle (cf. Sec. 13.4, page 177 of Ref. 73) is

$$Z = F(z) = \sqrt{m} \operatorname{sn}\left(\frac{2K}{\pi}\sin^{-1}\frac{z}{c}\right) \tag{7}$$

where $c^2 = a^2 - b^2$, and the modulus m is determined from the expression

$$\frac{K'}{K} = \frac{4}{\pi}\tanh^{-1}\frac{b}{a} = \frac{2}{\pi}\ln\frac{a+b}{a-b} \tag{8}$$

The derivative $F'(z)$ is found to be (cf. Ref. 12)

$$F'(z) = \frac{2K\sqrt{m}\operatorname{cn}u\operatorname{dn}u}{\pi\sqrt{c^2 - z^2}} \qquad u = \frac{2K}{\pi}\sin^{-1}\frac{z}{c} \tag{9}$$

Substituting these expressions into Eq. (6), we find for the discharge into a well (at z_1) with an elliptical contour

$$q_1 = \frac{2\pi(\phi_R - \phi_w)}{\ln(1 - k\operatorname{sn}^2 u_1) - \ln(\alpha|\operatorname{cn}u_1\operatorname{dn}u_1/\sqrt{c^2 - z_1^2}|)} \tag{10}$$

where
$$u_1 = \frac{2K}{\pi}\sin^{-1}\frac{z_1}{c} \qquad \alpha = \frac{2Kr_w\sqrt{m}}{\pi}$$

For a well at the center of the contour ($z_1 = 0$), Eq. (10) reduces to

$$q_1 = \frac{2\pi(\phi_R - \phi_w)}{\ln[\pi c/(2r_w K\sqrt{m})]} \tag{11}$$

Thus, if we form the ratio of this value of q_1 to that for strictly radial flow q_0, given by Eq. (8), Sec. 10-4, we obtain

$$\frac{q_1}{q_0} = \frac{\ln(R/r_w)}{\ln[\pi c/(2r_w K\sqrt{m})]} \tag{12}$$

To investigate the influence of the shape of the contour on the discharge, Polubarinova-Kochina [116] first considered an elliptical contour and a circular contour with the same area ($R^2 = ab$). Taking $a = nb$ and assuming a series of values of n for this case, she obtained curve I of

Fig. 10-6. Curve II was obtained by setting the minor semiaxis of the elliptical contour equal to the radius of the circular contour ($b = R$) and then taking a series of values of n with $a = nR$. It is seen from these curves that the discharge changes very little with changes in the form of the contour. In the latter case (curve II) if $n \rightarrow \infty$, the elliptical region

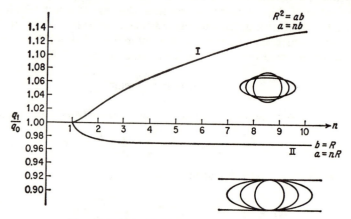

FIG. 10-6. (*After Polubarinova-Kochina* [116].)

deteriorates into an infinite strip, and the discharge is only 0.966 of that into the center of a circular contour with a diameter equal to the width of the strip.

10-6. Interference among Wells

Observations indicate that when a number of wells are introduced within a feed contour, the output increases, but the efficiency of each additional well decreases. This effect is due to the so-called *interference among wells*.

To demonstrate the phenomena of interference, consider two wells with the same discharge q located at points $z = \pm a$ within the boundary contour of the flow region. The complex potential for this case is then equal to the sum of the individual potentials, or

$$w = -\frac{q}{2\pi} [\ln (z - a) + \ln (z + a)] + C$$

$$w = -\frac{q}{2\pi} \ln [(z - a)(z + a)] + C \qquad (1)$$

The real part of Eq. (1) yields for the potential

$$\phi = -\frac{q}{2\pi} \ln |(z - a)(z + a)| + C_1 \qquad (2)$$

Assuming that each of the wells is of radius r_w and is the center of an equipotential ϕ_w at $z = a + r_w$, and considering $r_w{}^2$ to be of higher order (i.e., $a \gg r_w$), we get

$$\phi_w = -\frac{q}{2\pi} \ln 2ar_w + C_1 \tag{3}$$

Now, on the basis of the work of the previous sections, we shall assume that the boundary contour ϕ_R can be approximated by a circle with radius R. Hence, the potential at $r = R$, assuming R is sufficiently greater than a, is

$$\phi_R = -\frac{q}{2\pi} \ln R^2 + C_1 \tag{4}$$

Subtracting Eq. (3) from Eq. (4) and solving for the discharge, we obtain for each of the wells

$$q = \frac{2\pi(\phi_R - \phi_w)}{\ln (R^2/2ar_w)} \tag{5}$$

Comparing the quantity of flow into one well of a two-well system to that for a single well with simple radial flow q_0,

$$\frac{q}{q_0} = \frac{\ln (R/r_w)}{\ln [(R/r_w)(R/2a)]} \tag{6}$$

we see that whereas $R/2a > 1$ the quantity per well is less for a two-well system than for a single well. Also, we note in this expression that the efficiency of each of the wells increases as the distance between them ($2a$) increases; that is, the interference between wells is reduced as the spacing increases. In this regard, it should be noted that the discharge [Eq. (5)] of the wells is dependent on their mutual separation and not on their absolute positions.

Many additional arrangements of wells, such as triangular patterns, square patterns, batteries of wells in circular patterns, and linear arrays, can be found in the works of Muskat [99], Charny [18], Shchelkachev [133, 134], and others [61, 88, 112].

10-7. Partially Penetrating Well in Semi-infinite Media

In the previous sections the problem of fully penetrating wells was investigated, and the resulting equations for the discharge were obtained for a range of conditions. In practice, however, we often encounter wells that extend only part way through the bearing strata. In this and the next section we shall consider some aspects of these *partially penetrating wells*.

The simplest case of a partially penetrating well occurs when a well just penetrates the top surface of a semi-infinite porous medium (Fig.

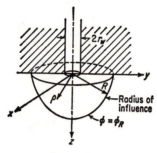

FIG. 10-7

10-7) and displays *spherical flow*. The discharge at any radial distance ρ from a source at the center of the sphere is

$$Q = 4\pi\rho^2 v_\rho = 4\pi\rho^2 \frac{\partial\phi}{\partial\rho} \tag{1}$$

where $4\pi\rho^2$ is the surface area of the sphere and v_ρ is the radial velocity. Hence the potential ϕ is

$$\phi = -\frac{Q}{4\pi\rho} + C \tag{2}$$

Equation (2) demonstrates that in the case of spherical flow the potential varies inversely with the radius. This is a more rapid variation than the logarithmic variation for the simple radial-flow problem. To determine the constant C in Eq. (2), we specify that at the well $\rho = r_w$ and $\phi = \phi_w$; hence $C = \phi_w + Q/4\pi r_w$ and

$$\phi = \frac{Q}{4\pi}\left(\frac{1}{r_w} - \frac{1}{\rho}\right) + \phi_w \tag{3}$$

Specifying that at $\rho = R$, $\phi = \phi_R$, and noting that $1/r_w \gg 1/R$, the discharge into the hemisphere (Fig. 10-7) is found to be

$$Q_s = 2\pi r_w(\phi_R - \phi_w) \tag{4}$$

Forming a ratio of the flows of a well producing by spherical flow (just penetrating the layer) to one producing by simple radial flow (completely penetrating a layer of thickness T), with the same potential drop,

$$\frac{Q_s}{Q_r} = \frac{r_w}{T}\ln\frac{R}{r_w} \tag{5}$$

we see that the efficiency of the spherical-flow system is dependent on the ratio r_w/T. Since, in general, the radius of a well is very much smaller than the thickness of the bearing layer, it is evident that the spherical-flow well is highly inefficient.   For example, if $r_w = \frac{1}{4}$ ft, $R/r_w = 1,000$, and $T = 50$ ft, the production of the radial-flow system will exceed the other by a factor of approximately 30.

An approximate solution to the problem of a well penetrating a semi-infinite porous mass to a depth of $L_0 = 2L$ (Fig. 10-8a) was obtained by

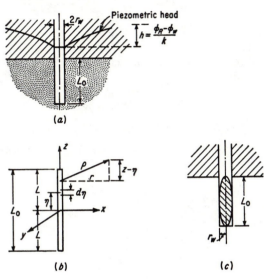

FIG. 10-8

Girinsky [44], assuming the discharge Q all along the well (Fig. 10-8b) to vary as

$$\frac{dQ}{d\eta} = \frac{Q}{2L} \tag{6}$$

Assuming Eq. (2) to represent the potential at any point in the flow region,

$$d\phi = \frac{dQ}{4\pi\rho} \tag{7}$$

Now, choosing the cylindrical coordinate system shown in Fig. 10-8b,

$$\rho = \sqrt{r^2 + (z - \eta)^2} \tag{8}$$

After combining Eqs. (6) and (7) and setting the limits of integration from $-L_0/2$ to $+L_0/2$, we find

$$\phi(r,z) = \frac{Q}{8\pi L} \int_{-L}^{L} \frac{d\eta}{\sqrt{r^2 + (z - \eta)^2}} = \frac{Q}{8\pi L} \left[\sinh^{-1}\frac{z + L}{r} - \sinh^{-1}\frac{z - L}{r} \right] \tag{9}$$

The equipotential surfaces given by Eq. (9) are seen to be ellipsoids. Assuming the average potential along the axis of the well to be that at $z = 0$ and substituting $L_0 = 2L$, Eq. (9) becomes

$$\phi(r) = \frac{Q}{2\pi L_0} \sinh^{-1} \frac{L_0}{2r} \approx \frac{Q}{2\pi L_0} \ln \frac{L_0}{r} \qquad (10)*$$

so that

$$Q = \frac{2\pi L_0 kh}{\ln (L_0/r_w)} \qquad (11)$$

where h is the head loss, as shown in Fig. 10-8a. As a correction to Eq. (11), Girinsky adjusted the major axis of the ellipsoid so that the volume of the ellipsoid would be equal to the volume of the cylinder (Fig. 10-8c), for which he obtained

$$Q = \frac{2\pi k L_0 h}{\ln (1.6 L_0/r_w)} \qquad (12)$$

The use of Eq. (12) will now be illustrated by an example.

Example 10-1. Determine the reduced discharge into the partially penetrating well shown in Fig. 10-9.

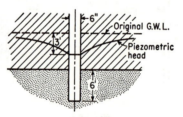

Fig. 10-9

Substituting into Eq. (12), $L_0 = 6$ ft, $h = 3$ ft, and $r_w = \frac14$ ft, we obtain $Q/k = 31$ ft.

10-8. Partially Penetrating Well in Layer of Finite Thickness

In Fig. 10-10a is shown a well penetrating a distance b into a layer of thickness T. Considering the discharge to be a constant over the entire surface of the well, Muskat [99] obtained the approximate formula

$$Q = \frac{2\pi T(\phi_R - \phi_w)}{(1/2\bar{T})[2 \ln (4T/r_w) - G(\bar{T})] - \ln (4T/R)} \qquad (1)$$

where

$$\bar{T} = \frac{b}{T}$$

$$G(\bar{T}) = \ln \frac{\Gamma(0.875\bar{T})\Gamma(0.125\bar{T})}{\Gamma(1 - 0.875\bar{T})\Gamma(1 - 0.125\bar{T})}$$

A plot of $G(\bar{T})$ as a function of $\bar{T}$ is shown in Fig. 10-10b.

* $\sinh^{-1} x = \ln (x + \sqrt{x^2 + 1}) \approx \ln 2x$ if $x \gg 1$.

Kozeny [77] obtained the simpler expression for Eq. (1):

$$Q = \frac{2\pi b(\phi_R - \phi_w)}{\ln(R/r_w)} \left(1 + 7\sqrt{\frac{r_w}{2b}} \cos\frac{\pi \bar{T}}{2}\right) \tag{2}$$

Recognizing the multiplier in Eq. (2) as the discharge for simple radial flow, we see that the second term in parenthesis is simply a correction factor for the flow from below. A plot of this correction factor is given in Fig. 10-10c.

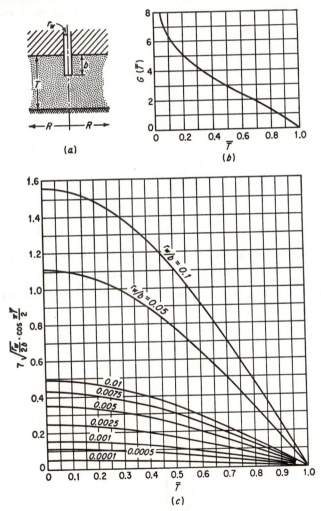

Fig. 10-10. (c) (*Based on Kozeny* [77].)

PROBLEMS

1. Find the radius of a single well to double the discharge from a well of radius r_w and radius of influence R, everything else being constant.

2. Demonstrate that Eqs. (5), Sec. 10-1, are conjugate harmonic functions.

3. A sink of strength q is located at the origin, and a source of $2q$ is at point a on the x axis. Determine the general expressions for ϕ and ψ.

4. Discuss the combination of the source-sink pattern of Sec. 10-3 with a uniform flow in the negative x direction.

5. If $R/r_w = 2,000$, determine the maximum displacement of a well from the center of a circular flow region such that the increase in the discharge will be less than 5 per cent.

6. Discuss the justification of assuming simple radial flow when the well bore is inclined to the plane of flow.

7. (a) Show that the complex potential for a well at point a of Fig. 10-11 is $w = q[\ln \sin (\pi/2L)(z - a) - \ln \sin (\pi/2L)(z + a)]/2\pi + C$. (b) Find the general expression for the discharge into the well at point a. (c) Establish the general expression for the discharge for the well at point a of Fig. 10-11 with $L \to \infty$.

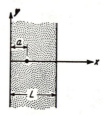

Fig. 10-11

8. Find the equivalent radius of a single well to produce the same discharge as a two-well system with the same head loss. Discuss the practicability of replacing two wells by an equivalent single well.

9. Establish Eq. (2), Sec. 10-7, directly from the statement of Laplace's equation in spherical coordinates.

Appendix A

A-1. Complex Numbers

Any number of the form $z = x + iy$, where x and y are scalars and $i = \sqrt{-1}$ is the *imaginary unit*, is called a *complex number*. The real number x is called the *real* part of z [written Re (z)] and the real number y is called the *imaginary* part of z [written Im (z)]. A complex number $z = x + iy$ can be taken as either the geometrical representation of a point with the cartesian coordinates (x,y) or as a position vector which radiates from the origin to this point (Fig. A-1). The cartesian plane, when used in this manner, is called

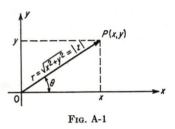

Fig. A-1

the *Argand diagram* or the *complex plane*, or simply the *z plane*. The magnitude of the vector OP and hence also the magnitude of z is given by

$$r = \sqrt{x^2 + y^2} = |x + iy| = |z| \tag{1}$$

and is called the *absolute* value or the *modulus* of z (written mod z). The direction angle θ, defined as

$$\theta = \tan^{-1}\frac{y}{x} \tag{2}$$

is called the *amplitude* or *argument* of z (written arg z).

Noting that $x = r \cos \theta$ and $y = r \sin \theta$, we obtain the trigonometric representation of a complex number

$$z = r(\cos \theta + i \sin \theta) \tag{3}$$

Now recalling from the calculus that

$$e^x = 1 + x + \frac{x^2}{2!} + \frac{x^3}{3!} + \cdots$$

$$e^{i\theta} = 1 + i\theta - \frac{\theta^2}{2!} - \frac{i\theta^3}{3!} + \cdots = \left(1 - \frac{\theta^2}{2!} + \frac{\theta^4}{4!} - \cdots\right)$$
$$+ i\left(\theta - \frac{\theta^3}{3!} + \frac{\theta^5}{5!} - \cdots\right)$$

we see that

$$e^{i\theta} = \cos\theta + i\sin\theta \qquad (4)$$

and
$$z = x + iy = r(\cos\theta + i\sin\theta) = re^{i\theta} \qquad (5)$$

Equation (5) demonstrates that the complex number $z = x + iy$ can also be expressed in an equivalent polar form. This result, first shown by Euler, is of great importance. Noting that $i = \sqrt{-1}$ can be written as the complex number $\exp(i\pi/2)$, which has a modulus of unity and an argument of $\pi/2$, we see that the multiplication of a complex number by i simply rotates the position vector representing the number through the angle $\pi/2$ without altering its modulus. For example, if $z = r \exp i\theta$, then $iz = ir \exp i\theta = r \exp[i(\theta + \pi/2)]$.

Since the symbolic statement of a complex number $z = x + iy$ demonstrates that the y axis is normal to the x axis, it follows that two complex numbers can be equal if and only if their respective real and imaginary parts are equal. That is, if $a + ib = c + id$, then $a = c$ and $b = d$. The extension of this concept to the addition and subtraction of two or more complex numbers is obvious.

Example 1. If $z_1 = 4 + 3i$ and $z_2 = 3 + 4i$, then

$$z_1 + z_2 = 7 + 7i$$
and
$$z_1 - z_2 = 1 - i$$

Multiplication and division of complex numbers follow immediately from Eq. (5) and the familiar rules of exponentials. Thus, if

$$z_1 = r_1 \exp i\theta_1$$

and $z_2 = r_2 \exp i\theta_2$ are two complex numbers, their product will be

$$z_1 z_2 = r_1 r_2 \exp[i(\theta_1 + \theta_2)]$$

and their quotient $z_1/z_2 = (r_1/r_2) \exp[i(\theta_1 - \theta_2)]$. Also from Eq. (5) it is evident that

$$z^n = (re^{i\theta})^n = r^n(\cos\theta + i\sin\theta)^n = r^n(\cos n\theta + i\sin n\theta)$$

and if $r = 1$, we have *de Moivre's theorem,*

$$(\cos\theta + i\sin\theta)^n = \cos n\theta + i\sin n\theta \qquad (6)$$

Example 2. Determine the double-angle formulas for the sine and cosine. Letting $n = 2$ in Eq. (6), we have

$$\cos^2\theta + 2i\sin\theta\cos\theta - \sin^2\theta = \cos 2\theta + i\sin 2\theta$$

which, separating real and imaginary parts, yields

$$\cos 2\theta = \cos^2\theta - \sin^2\theta \qquad \sin 2\theta = 2\sin\theta\cos\theta$$

Equation (6) can readily be extended to obtain integral roots of z. Let

$$w = R(\cos \phi + i \sin \phi) = z^{1/n}$$

Then $\quad w^n = R^n(\cos n\phi + i \sin n\phi) = z = r(\cos \theta + i \sin \theta)$

Since two complex numbers which are equal must have the same modulus, $R^n = r$ or $R = r^{1/n}$, which require only a real arithmetic calculation. Also, the arguments of equal complex numbers must be equal or differ by an integral multiple of 2π. Thus, $n\phi = \theta + 2\pi k$ or $\phi = (\theta + 2\pi k)/n$, where $k = 0, 1, 2, \ldots, (n - 1)$; and the n roots are

$$w = r^{1/n}\left(\cos \frac{\theta + 2k\pi}{n} + i \sin \frac{\theta + 2k\pi}{n}\right) \tag{7}$$

Equation (7) demonstrates that the n roots are located on the circumference of a circle with its center at the origin and radius $r^{1/n}$; they are spaced at equal angular intervals of $2\pi/n$ from the point on the circle subtended by the angle θ/n.

Example 3. Determine all values of $i^{\frac{1}{3}}$.

The modulus of i equals unity and its argument is $\pi/2$, therefore

$$\sqrt[3]{i} = \sqrt[3]{1\left(\cos \frac{\pi}{2} + i \sin \frac{\pi}{2}\right)}$$
$$= \left(\cos \frac{\pi/2 + 2k\pi}{3} + i \sin \frac{\pi/2 + 2k\pi}{3}\right) \qquad k = 0, 1, 2$$

and the three roots are (Fig. A-2):

$$\cos \frac{\pi}{6} + i \sin \frac{\pi}{6} = \frac{\sqrt{3}}{2} + \frac{i}{2}$$
$$\cos \frac{5\pi}{6} + i \sin \frac{5\pi}{6} = -\frac{\sqrt{3}}{2} + \frac{i}{2}$$
$$\cos \frac{3\pi}{2} + i \sin \frac{3\pi}{2} = -i$$

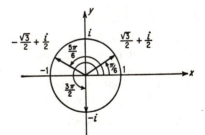

Fig. A-2

The roots are shown by the position vectors on the circumference of the unit circle $[(\bmod i)^{\frac{1}{3}} = 1]$. Following the procedure outlined above, the first root is located by the vector at the direction angle of $\pi/6$ $[\theta/n]$, and the subsequent roots are spaced at equal intervals of $2\pi/3$ $[2\pi/n]$.

Let us now consider the natural logarithm of a complex number, $\ln z$. Employing the exponential form, we have

$$\ln z = \ln re^{i\theta} = \ln r + i\theta \tag{8}$$

or, recalling that $\theta = \tan^{-1} (y/x) = \arg z$,

$$\ln z = \ln |z| + i \arg z \qquad (9)*$$

Example 4. Determine the principal value of $\ln (4 - 3i)$.

The modulus is $(4^2 + 3^2)^{1/2} = 5$ and the argument is $\tan^{-1} (-\frac{3}{4}) = 5.46$ radians. Therefore

$$\ln (4 - 3i) = \ln 5 + 5.46i$$
$$= 1.61 + 5.46i$$

RECOMMENDED EXERCISES

1. Reduce the following to the form $a + ib$:

(a) $(1 + i)^2 + (3 + 2i)^2$ (b) $\dfrac{1 - 3i}{2 + i}$

(c) $\dfrac{4 + 3i}{i}$

2. Obtain in trigonometric form the complex numbers:

(a) $2 \sqrt{3} + 2i$ (b) $2 \sqrt{3} - 2i$
(c) $3i$ (d) -2

3. Write each of the complex numbers in Exercise 2 in exponential form.
4. What is the effect of multiplying a complex number by $i^{1/2}$? $i^{3/2}$?
5. Verify the following identities:

(a) $\cos 3\alpha = \cos^3 \alpha - 3 \cos \alpha \sin^2 \alpha$ (b) $\sin 3\alpha = 3 \cos^2 \alpha \sin \alpha - \sin^3 \alpha$

6. Find all the cube roots of $(1 + i)$.
7. Find $\ln (-1)$, $\ln (-i)$, $\ln (3 + 2i)$.

A-2. Absolute Values

In Eq. (1) we defined $|z|$ as the absolute value or modulus of z and noted that it designated either the length of the position vector z or the distance of the point z from the origin. Hence, following the same definition, it follows that $|z_1 - z_2|$ is the distance between the two points z_1 and z_2 in the complex plane, since

$$|z_1 - z_2| = |(x_1 - x_2) + i(y_1 - y_2)| = \sqrt{(x_1 - x_2)^2 + (y_1 - y_2)^2}$$

Likewise the statement $|z - i| = 2$ requires the point z to lie on a circle of radius 2 with a center at the point $(0,i)$.

The conjugate of a complex number is defined as the reflection of the

---

* If θ_1 is the *principal argument* of z, that is, $0 \leq \theta_1 \leq 2\pi$, Eq. (9) can be written as

$$\ln z = \ln |z| + i(\theta_1 + 2\pi n) \qquad n = 0, \pm 1, \pm 2, \ldots$$

which demonstrates that the logarithmic function is infinitely many-valued. If $n = 0$, the value of the function is called the *principal value* of $\ln z$.

number in the x axis (Fig. A-3). For example, the conjugate of $z = x + iy$ is the complex number $\bar{z} = x - iy$. The conjugate of a complex number

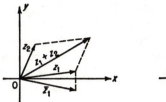

FIG. A-3

is denoted by placing a bar over the number. Obviously,

$$|\bar{z}| = |z|$$

From the properties of a triangle (Fig. A-3), it is evident that

$$|z_1 + z_2| \leqq |z_1| + |z_2| \tag{10a}$$
$$|z_1 - z_2| \geqq ||z_1| - |z_2|| \tag{10b}$$

or, in general

$$\left| \sum_{k=1}^{n} z_k \right| \leqq \sum_{k=1}^{n} |z_k| \qquad n = 1, 2, 3, \ldots$$

RECOMMENDED EXERCISES

1. Obtain an algebraic expression in x and y, and define the region in the z plane for

(a) $\arg z = \frac{\pi}{2}$

(b) $|z| < 2$

(c) $|z - 5| < 10$

(d) $|z - 1| \geqq 1$

(e) $0 < \text{Re}(z) \leqq \text{Im}(z)$

(f) $|z - 2| = |z - 2i|$

(g) $\frac{1}{z} > 2$

(h) $|z - 2| \leqq \text{Im}(z)$

2. Demonstrate:

(a) $|z_1 - z_2| \leqq |z_1| + |z_2|$

(b) $|z_1 + z_2| \geqq ||z_1| - |z_2||$

A-3. Analytic Functions

When x and y in the statement $z = x + iy$ are variables, z is called a complex variable. If to each value of z there corresponds a second complex variable w, then w is said to be a function of the complex variable z; that is, $w = f(z)$, such as, $w = z^2$, $w = \cos z$, and $w = \ln z$. In particular, let us take the complex variable w as

$$w = \phi + i\psi \tag{11}$$

wherein the dependence of ϕ and ψ on x and y is implied. For example, if $w = z^2$, $\phi = x^2 - y^2$ and $\psi = 2xy$.

In groundwater problems we shall generally restrict w to a class of functions known as *analytic functions*, sometimes also called *regular* or *holomorphic functions*. If, at each point within a simple closed curve or *contour* (also *domain* or *region*) of the z plane, the function $w = f(z)$ and its derivative dw/dz are both single-valued and finite, then the function is said to be analytic within the contour. Generally it will be sufficient to consider a contour as a closed curve (formed from a continuous chain

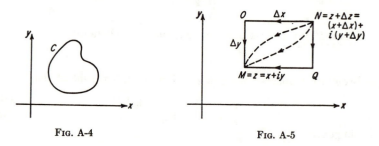

FIG. A-4   FIG. A-5

of arcs) which does not intersect itself (Fig. A-4). The requirement that a function be single-valued and finite will be satisfied if there corresponds one and only one value of w to each value of z within the contour, and if its modulus is finite.

The necessary condition that the derivative dw/dz be single-valued and finite within the contour is provided by the Cauchy-Riemann equations which we shall now develop.

The derivative of a function of a complex variable $w = f(z) = \phi + i\psi$ is defined to be

$$\frac{dw}{dz} = f'(z) = \lim_{\Delta z \to 0} \frac{f(z + \Delta z) - f(z)}{\Delta z} \qquad (12)$$

This definition is entirely similar to that for functions of real variables; that is, if $w = z^n$, $f'(z) = nz^{n-1}$. However, unlike real variables,

$$\Delta z = \Delta x + i \, \Delta y$$

is itself a complex variable which can approach zero along infinitely many paths. To illustrate, in Fig. A-5 the point $N = z + \Delta z$ can approach the point $M = z$ along any of the paths shown. In particular, let us consider two cases: (1) N tends to M parallel to the x axis (along NO), thus, $\Delta y = 0$ and $\Delta z = \Delta x$; (2) N tends to M parallel to the y axis (along NQ), thus,

$\Delta x = 0$ and $\Delta z = i\,\Delta y$. In the general case, Eq. (12) becomes

$$\frac{dw}{dz} = \lim_{\substack{\Delta x \to 0 \\ \Delta y \to 0}} \frac{\phi(x + \Delta x,\, y + \Delta y) - \phi(x,y) + i[\psi(x + \Delta x,\, y + \Delta y) - \psi(x,y)]}{\Delta x + i\,\Delta y} \tag{13}$$

For case (1),

$$\frac{dw}{dz} = \lim_{\Delta x \to 0} \left[\frac{\phi(x + \Delta x,\, y) - \phi(x,y)}{\Delta x} + i\,\frac{\psi(x + \Delta x,\, y) - \psi(x,y)}{\Delta x} \right] \tag{14a}$$

and for case (2),

$$\frac{dw}{dz} = \lim_{\Delta y \to 0} \frac{1}{i} \left[\frac{\phi(x,\, y + \Delta y) - \phi(x,y)}{\Delta y} + i\,\frac{\psi(x,\, y + \Delta y) - \psi(x,y)}{\Delta y} \right] \tag{14b}$$

Recognizing that the difference quotients in Eqs. (14) are partial derivatives with respect to x and y, we have, respectively,

$$\frac{dw}{dz} = \frac{\partial \phi}{\partial x} + \frac{i\,\partial \psi}{\partial x} \tag{15a}$$

and

$$\frac{dw}{dz} = \frac{\partial \psi}{\partial y} - \frac{i\,\partial \phi}{\partial y} \tag{15b}$$

Now, if the derivative dw/dz is to be single-valued, it is necessary that Eqs. (15) be equal. Hence

$$\frac{\partial \phi}{\partial x} = \frac{\partial \psi}{\partial y} \tag{16a}$$

$$\frac{\partial \psi}{\partial x} = -\frac{\partial \phi}{\partial y} \tag{16b}$$

These two extremely important conditions are called the Cauchy-Riemann equations. They are necessary but not sufficient. However, it can be proved [20] that the requirement of sufficiency is completely satisfied if all the partial derivatives $\partial \phi/\partial x$, $\partial \phi/\partial y$, $\partial \psi/\partial x$, and $\partial \psi/\partial y$ are continuous within the region. Thus, in summary, if ϕ and ψ have continuous derivatives of the first order in the region C, a necessary and sufficient condition that $w = \phi + i\psi$ be an analytic function of $z = x + iy$ in C is that the Cauchy-Riemann equations are satisfied. If a function $f(z)$ is not analytic at some point z_0 but is analytic at every point in the neighborhood of z_0, then z_0 is said to be a *singular point* or a *singularity* of the function. For example, if $f(z) = (z - a)^{-1}$ the function is analytic at every point except the point $z = a$, where it is discontinuous; hence $z = a$ is a singular point.

Although such simple functions as $\bar{z}$ and $|z|$ are nowhere analytic, it will be shown that if a function is analytic it possesses derivatives, not only of the first order, but of all orders. Thus the existence and continuity of all partial derivatives of ϕ and ψ are assured if $w = f(z)$ is analytic.

Of the many important properties of analytic functions we note the following:

Property 1. Both the real (ϕ) and imaginary (ψ) parts of an analytic function satisfy Laplace's equation in two dimensions,

$$\frac{\partial^2\theta}{\partial x^2} + \frac{\partial^2\theta}{\partial y^2} = 0 \tag{17a}$$

Differentiating the first of Eqs. (16) with respect to x and the second with respect to y and adding, we obtain

$$\frac{\partial^2\phi}{\partial x^2} + \frac{\partial^2\phi}{\partial y^2} = 0 \tag{17b}$$

In a similar manner for ψ, changing the order of differentiation,

$$\frac{\partial^2\psi}{\partial x^2} + \frac{\partial^2\psi}{\partial y^2} = 0 \tag{17c}$$

Any function which satisfies this equation is called an *harmonic function*. Two harmonic functions such as ϕ and ψ, so related that $\phi + i\psi$ is an analytic function, are called *conjugate harmonic functions*.

Property 2. If ϕ and ψ are conjugate harmonic functions, then the families of curves $\phi(x,y) = c$ and $\psi(x,y) = k$, where c and k are constants, intersect each other at right angles.

To prove this we obtain the slope of each family of curves, getting, respectively,

$$\frac{dy}{dx} = -\frac{\partial\phi/\partial x}{\partial\phi/\partial y} \tag{18a}$$

and

$$\frac{dy}{dx} = -\frac{\partial\psi/\partial x}{\partial\psi/\partial y} \tag{18b}$$

Since $w = \phi + i\psi$ is analytic, ϕ and ψ must satisfy the Cauchy-Riemann equations. Hence, substituting Eqs. (16) into Eq. (18b), we obtain

$$\frac{dy}{dx} = \frac{\partial\phi/\partial y}{\partial\phi/\partial x}$$

which is the negative reciprocal of the slope of the curves $\phi(x,y) = c$ [Eq. (18a)].

RECOMMENDED EXERCISES

1. If $w = \phi + i\psi = f(z)$, find ϕ and ψ for the following and determine where the function is analytic.

(a) $w = \dfrac{1}{z}$ (b) $w = e^z$

(c) $w = \ln z$ (d) $w = z - 1$

(e) $w = \cos z$ (f) $w = \cos x$

2. At what points within the region $|z - 1| < 2$ does the function $(z - 5)/(z - 6)(z^2 - 4)$ fail to be analytic?

3. Choose the real constants a, b, and c so that the function $w = x + 2ay + i(bx + cy)$ is analytic.

A-4. Elementary Functions of z; Hyperbolic Functions

From Eq. (4), written as

$$e^{iz} = \cos z + i \sin z$$

and its companion equation,

$$e^{-iz} = \cos z - i \sin z$$

we obtain, by addition and subtraction, the definition of the sine and cosine of a complex variable z:

$$\cos z = \frac{e^{iz} + e^{-iz}}{2} \tag{19a}$$

$$\sin z = \frac{e^{iz} - e^{-iz}}{2i} \tag{19b}$$

Similar in form to Eqs. (19) are the hyperbolic cosine and sine, defined as

$$\cosh z = \frac{e^z + e^{-z}}{2} \tag{20a}$$

$$\sinh z = \frac{e^z - e^{-z}}{2} \tag{20b}$$

The hyperbolic tangent of z is defined by the equation

$$\tanh z = \frac{\sinh z}{\cosh z} \tag{20c}$$

and the associated functions coth z, sech z, and csch z are defined, respectively, as the reciprocals of tanh z, cosh z, and sinh z. The hyperbolic functions are widely tabulated (cf. Refs. 29, 111).

From Eqs. (19) and (20) it follows that:

$$\sinh{(iz)} = i \sin z$$
$$\sin{(iz)} = i \sinh z$$
$$\cosh{(iz)} = \cos z$$
$$\cos{(iz)} = \cosh z$$
$$\cos z = \cos x \cosh y - i \sin x \sinh y \qquad (21)$$
$$\sin z = \sin x \cosh y + i \cos x \sinh y$$
$$\cosh z = \cosh x \cos y + i \sinh x \sin y$$
$$\sinh z = \sinh x \cos y + i \cosh x \sin y$$
$$\cosh^2 z - \sinh^2 z = 1$$

Some of the more important characteristics of hyperbolic functions are the relationships between the inverses of these functions and certain logarithmic expressions. For example, if we take $w = \cosh^{-1} z$,

$$z = \cosh w = \frac{e^w + e^{-w}}{2}$$

and

$$(e^w)^2 - 2ze^w + 1 = 0$$

Solving this quadratic for e^w, we obtain

$$e^w = z \pm \sqrt{z^2 - 1}$$

which, taking the natural logarithm of both sides and noting that $w = \cosh^{-1} z$, yields

$$\cosh^{-1} z = \ln{(z \pm \sqrt{z^2 - 1})} \qquad (22a)$$

Similarly, it can be shown that

$$\sinh^{-1} z = \ln{(z + \sqrt{z^2 + 1})} \qquad (22b)$$
$$\tanh^{-1} z = \frac{1}{2} \ln \frac{1 + z}{1 - z} \qquad (22c)$$
$$\coth^{-1} z = \frac{1}{2} \ln \frac{z + 1}{z - 1} \qquad (22d)$$

RECOMMENDED EXERCISES

1. Verify each of the expressions in Eqs. (21) and (22).
2. Find the logarithmic equivalents for $\sin^{-1} z$ and $\tan^{-1} z$.
3. Evaluate

(a) $\cos{(1 + 2i)}$ (b) $\tan^{-1} i$
(c) $\cosh^{-1}{(-2)}$ (d) $\tanh^{-1} \frac{1}{2}$
(e) $\cos^{-1} 2$

A-5. Complex Integration

Let $f(z)$ be a continuous function of the complex variable z on the arc of a curve AB (Fig. A-6). Consider AB to be divided into n parts, corresponding to the sections of arc between the sequence of points

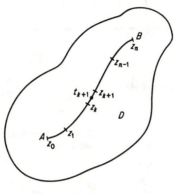

FIG. A-6

$A = z_0, z_1, \ldots, z_k, z_{k+1}, \ldots, z_{n-1}$, and $z_n = B$. Further, let t_k denote an intermediate point in each interval of arc $z_{k-1}z_k$. Calling the interval $\Delta z_k = z_k - z_{k-1}$, we form the sum

$$\sum_{k=1}^{n} f(t_k)\, \Delta z_k$$

Now, if the length of each interval approaches zero ($\Delta z_k \to 0$) while their number approaches infinity ($n \to \infty$), the limit of the summation above is called the line integral of $f(z)$ along AB; that is,

$$\int_{AB} f(z)\, dz = \lim_{\substack{n \to \infty \\ \Delta z_k \to 0}} \sum_{k=1}^{n} f(t_k)\, \Delta z_k \qquad (23a)$$

Writing $f(z) = \phi(x,y) + i\psi(x,y)$ and noting that $dz = dx + i\, dy$, we obtain from Eq. (23a)

$$\int_{AB} f(z)\, dz = \int_{AB} (\phi\, dx - \psi\, dy) + i \int_{AB} (\psi\, dx + \phi\, dy) \qquad (23b)$$

Example 5. Find the line integral of the function $f(z) = z^2$ along the path from A through B to C in Fig. A-7.

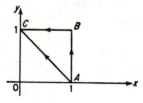

Fig. A-7

Here $f(z) = \phi + i\psi = z^2$; therefore

$$\phi = x^2 - y^2 \qquad \psi = 2xy$$

Along the vertical path AB, $dx = 0$ and $x = 1$, whereas, along the horizontal path BC, $dy = 0$ and $y = 1$. Hence Eq. (23b) becomes

$$I_1 = \int_A^C z^2\, dz = \int_A^B z^2\, dz + \int_B^C z^2\, dz$$
$$= -\int_0^1 2xy\, dy\,\Big|_{x=1} + i\int_0^1 (x^2 - y^2)\, dy\,\Big|_{x=1} + \int_1^0 (x^2 - y^2)\, dx\,\Big|_{y=1}$$
$$+ i\int_1^0 2xy\, dx\,\Big|_{y=1}$$
$$= -\tfrac{1}{3}(1 + i)$$

If the path of integration was along AOC, we would have

$$I_2 = \int_A^C z^2\, dz = \int_1^0 x^2\, dx - i\int_0^1 y^2\, dy$$
$$= -\tfrac{1}{3}(1 + i)$$

We note that $I_1 = I_2$. Hence if we were to perform the integration unidirectionally over the closed contour $ABCOA$, we would obtain $I_2 - I_1 = 0$ or $\int_{ABCOA} z^2\, dz = 0$. This is a consequence of the integrand $f(z) = z^2$ being analytic within and on the contour of $ABCOA$. We shall now state this as a theorem, known as *Cauchy's integral theorem*.

Theorem. If C is a closed curve interior to a region R within and on which $f(z)$ is analytic, then

$$\oint_C f(z)\, dz = 0 \qquad (23c)^*$$

* The symbol $\oint_C$ denotes the contour integration around the curve C in the direction of the arrow; that is, so that the region within the contour is always to the left of an observer moving in this direction along C. In general $\oint_C = -\oint_{C'}$.

The proof of this theorem can be found in texts on complex variable theory (cf. Refs. 20, 67).

An important corollary of Cauchy's integral theorem which was demonstrated in Example 5 is: if $f(z)$ is analytic within and on a region R, then the value of the line integral between any two points within R is independent of the path of integration within R. For example, in Fig. A-7, since $f(z) = z^2$ is analytic in and on $ABCOA$,

$$\oint_{ABCOA} z^2 \, dz = 0 = \int_{ABC} + \int_{COA} = \int_{ABC} - \int_{AOC}$$

and

$$\int_{ABC} = \int_{AOC}$$

Example 6. Find the line integral of the function $f(z) = z^2$ along the straight-line path AC in Fig. A-7.

Whereas $y = x - 1$ along this path

$$\phi = x^2 - y^2 = 2x - 1 \qquad \text{or} \qquad \phi = 1 - 2y$$
$$\psi = 2xy = 2x - 2x^2 \qquad \text{or} \qquad \psi = 2y - 2y^2$$

Substituting into Eq. (23b),

$$\int_A^C z^2 \, dz = \int_1^0 (2x - 1) \, dx - \int_0^1 (2y - 2y^2) \, dy + i \int_1^0 (2x - 2x^2) \, dx$$
$$+ i \int_0^1 (1 - 2y) \, dy$$

$$= -\tfrac{1}{3}(1 + i)$$

which is precisely the same value obtained in Example 5 for the other paths.

In the doubly connected region shown in Fig. A-8, $f(z)$ is assumed to be analytic throughout the region between and on the contours C_1 and C_2;

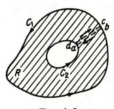

Fig. A-8

however, no restrictions are set on $f(z)$ interior to contour C_2. Introducing infinitesimal cuts shown by the dotted lines ab and cd and applying Cauchy's integral theorem, we have

$$\oint_{C_1} f(z) \, dz + \int_a^b f(z) \, dz - \oint_{C_2} f(z) \, dz + \int_c^d f(z) \, dz = 0$$

Now, noting that the second and fourth integrals are equal and opposite and hence cancel, we have

$$\oint_{C_1} f(z)\,dz = \oint_{C_2} f(z)\,dz \qquad (24a)$$

Equation (24a) demonstrates the *principle of the contraction (or of enlargement) of a contour:* The contour integral of an analytic function around

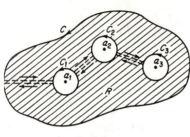

any curve can be replaced by the contour integral of the same function with the same directional sense* around any other curve into which the first can be continuously contracted (or enlarged) provided that in doing so it does not pass over a point where the function $f(z)$ fails to be analytic. This principle can be extended to regions of any order of connectivity. For example, in Fig.

FIG. A-9

A-9 if we consider each of the small circles shown to be drawn around isolated singular points, then by the same procedure outlined above, we obtain

$$\oint_C f(z)\,dz = \oint_{C_1} f(z)\,dz + \oint_{C_2} f(z)\,dz + \oint_{C_3} f(z)\,dz \qquad (24b)$$

and the integral around a contour is seen to be equal to the sum of the integrals around small circles with centers at isolated singular points $(a_1, a_2, \ldots, a_n)$.

RECOMMENDED EXERCISES

1. In Fig. A-7, if $f(z) = |z^2|$, obtain the line integral of $f(z)$ from A to C along the path ABC.

2. Integrate $f(z) = \sin z$ along each of the paths in Fig. A-7.

A-6. Integration around Singular Points; Residues

If a function $f(z)$ is analytic within and on a closed contour C of a simply connected region R, and if point a is interior to C, then

$$f(a) = \frac{1}{2\pi i} \oint_C \frac{f(z)\,dz}{z - a} \qquad (25a)$$

Equation (25a) is *Cauchy's integral formula.* It shows that the value of

---

* That is, if the direction of integration around the contour C_1 in Fig. A-8 is such that the interior region is to the left of the observer making the traverse, then the same must hold true for the observer traveling around the contour C_2.

a function that is analytic within a region is completely determined throughout the region in terms of its values on the boundary.

Example 7. If C is the circle $|z - 1| = 1$, evaluate the integral

$$\oint_C \frac{2z^2 + 3}{z^2 - 1}\, dz$$

We take the integral as

$$\oint_C \frac{2z^2 + 3}{z + 1}\, \frac{dz}{z - 1}$$

The function $f(z) = (2z^2 + 3)/(z + 1)$ is seen to be analytic at all points within C. Thus, applying Cauchy's integral formula with $a = 1$, we have

$$\oint_C \frac{2z^2 + 3}{z + 1}\, \frac{dz}{z - 1} = 2\pi i f(1) = 2\pi i\, \frac{2 + 3}{1 + 1} = 5\pi i$$

Derivatives of any order of $f(z)$ may be obtained by differentiating the right-hand side of Eq. (25a) with respect to a:

$$f'(a) = \frac{1}{2\pi i} \oint_C \frac{f(z)}{(z - a)^2}\, dz$$

In general,

$$f^{(n)}(a) = \frac{n!}{2\pi i} \oint_C \frac{f(z)}{(z - a)^{n+1}}\, dz \tag{25b}$$

Equation (25b) demonstrates that analytic functions possess derivatives of all orders, and hence the derivative of any order of any analytic function $f^{(n)}(z)$ is also analytic.

As a consequence of Cauchy's integral formula, if $f(z)$ is analytic within and on a circle $|z - a| = r$, then for every z interior to the circle the *Taylor's series*

$$f(z) = f(a) + \sum_{n=1}^{\infty} \frac{f^{(n)}(a)}{n!}\, (z - a)^n \tag{26}$$

is a valid representation of the function. The radius of the largest circle which can be drawn with $z = a$ as a center that contains only points where $f(z)$ is analytic is said to be the *radius of convergence* of the series. If $a = 0$ in Eq. (26), the resulting series is *Maclaurin's series*.

Example 8. Expand $\ln z$ into a Taylor's series about $z = 1$.
Noting

$$f(z) = \ln z,\, f'(z) = \frac{1}{z},\, f''(z) = -\frac{1}{z^2}, \cdots, f^{(n)}(z) = (-1)^{n-1}\, \frac{(n - 1)!}{z^n}$$

we obtain for the coefficients, with $z = 1$,

$$f(1) = 0,\, f'(1) = 1,\, \frac{f''(1)}{2!} = -\frac{1}{2}, \cdots, \frac{f^{(n)}(1)}{n!} = \frac{(-1)^{n-1}}{n}$$

and consequently the required expansion is

$$\ln z = z - 1 - \frac{1}{2}(z-1)^2 + \cdots + \frac{(-1)^{n-1}}{n}(z-1)^n - \cdots$$

If $f(z)$ is analytic within the annular region $r < |z - a| < R$ (Fig. A-10) its representation within this region is given by the *Laurent series*

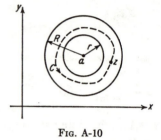

Fig. A-10

$$f(z) = \sum_{n-\infty}^{\infty} A_n(z-a)^n \qquad (27a)$$

where, if C is any closed curve lying between r and R (Fig. A-10),

$$A_n = \frac{1}{2\pi i} \oint_C \frac{f(z)}{(z-a)^{n+1}} \, dz$$
$$n = 0, \pm 1, \pm 2, \ldots \quad (27b)$$

The coefficient of the $(z-a)^{-1}$ term in the expansion,

$$A_{-1} = \frac{1}{2\pi i} \oint_C f(z) \, dz \qquad (27c)$$

is called the *residue of $f(z)$* at the isolated singular point a and is of particular importance.

If the Laurent expansion of $f(z)$ around an isolated singular point contains an infinite number of nonvanishing negative powers of $(z - a)$, then $z = a$ is called an *essential singular point* of $f(z)$. If the Laurent expansion contains only a finite number of negative powers of $(z - a)$ up to the nth term, then the function is said to have a *pole of order n at $z = a$*. In particular, a pole of order $n = 1$ is called a *simple pole*. Thus, for example, the function

$$e^{1/z} = 1 + \frac{1}{z} + \frac{1}{2!z^2} + \frac{1}{3!z^3} + \cdots$$

has an essential singularity at $z = 0$, whereas the function

$$\cot z = \frac{\cos z}{\sin z} = \frac{1}{z} - \frac{z}{3} - \frac{z^3}{45} - \cdots$$

possesses a simple pole at $z = 0$. Note that the coefficient of the z^{-1} term in the latter series, and hence the residue of the function, is $A_{-1} = 1$.

To expand upon the significance of the residue of a function, consider the integral

$$\oint_C (z-a)^n \, dz$$

where C is a circle of radius r and center a, and n is an integer. Making the substitutions $z - a = re^{i\theta}$ and $dz = rie^{i\theta}\,d\theta$, the integral becomes

$$r^{n+1} i \int_0^{2\pi} e^{(n+1)i\theta}\,d\theta = \frac{r^{n+1}}{n+1} e^{(n+1)i\theta}\Big|_0^{2\pi} = \begin{cases} 0 & n \neq -1 \\ 2\pi i & n = -1 \end{cases} \quad (28)$$

Now, taking the Laurent expansion of $f(z)$ in the neighborhood of $z = a$ [Eq. (27a)],

$$f(z) = \cdots + A_2(z - a)^2 + A_1(z - a) + A_0 + \frac{A_{-1}}{z - a} + \frac{A_{-2}}{(z - a)^2} + \cdots$$

and integrating around the circle C, from the results of Eq. (28), we obtain

$$\oint_C f(z)\cdot dz = 2\pi i A_{-1}$$

which is precisely the expression given in Eq. (27c). Thus we see that $2\pi i$ times the residue of a function represents the value of the integral of the function around a region within which the function is analytic at all points except for an isolated singularity.

Example 9. If C is the circle $|z| < \pi$, evaluate the integral

$$\oint_C \cot z\,dz$$

Noting from above that the residue of $\cot z$ is $A_{-1} = 1$, we obtain immediately

$$\oint_C \cot z\,dz = 2\pi i(A_{-1}) = 2\pi i$$

The extension of the residue theory to a finite number of isolated singular points follows directly from Eq. (24b). If a function $f(z)$ is analytic within and on a simple closed curve C (Fig. A-9), except for a finite number of isolated singular points interior to C, then

$$\oint_C f(z)\,dz = 2\pi i(\alpha_1 + \alpha_2 + \cdots + \alpha_n) \quad (29)$$

where $\alpha_1, \alpha_2, \ldots, \alpha_n$ are the residues of $f(z)$ at each of the isolated singular points within C.

Several methods are available for the determination of the residue of a function. However, because of the nature of the problems dealt with in the text the following method is particularly advantageous.

Let $z = a$ be an nth order pole of $f(z)$. Then

$$f(z) = \frac{A_{-n}}{(z - a)^n} + \cdots + \frac{A_{-1}}{z - a} + \phi(z)$$

where $\phi(z)$ is analytic in the circle about point a. Multiplying this identity by $(z - a)^n$ and differentiating with respect to z, $(n - 1)$ times, we obtain

$$\frac{d^{n-1}}{dz^{n-1}}[(z - a)^n f(z)] = (n - 1)!A_{-1} + \frac{d^{n-1}}{dz^{n-1}}[(z - a)^n \phi(z)]$$

Now, if we let $z \to a$, we obtain for the residue

$$A_{-1} = \frac{1}{(n - 1)!} \lim_{z \to a} \frac{d^{n-1}}{dz^{n-1}}[(z - a)^n f(z)] \tag{30a}$$

In particular, if $z = a$ is a simple pole,

$$A_{-1} = \lim_{z \to a}[(z - a)f(z)] \tag{30b}$$

Example 10. Evaluate the integral

$$\oint_C \frac{z}{z^2 - 1} dz$$

where C is the circle $|z - \frac{1}{2}| = 1$.

In this case although there are two isolated singular points, $z = \pm 1$, only $z = 1$ lies within the curve C. Hence, for the residue of the function at $z = 1$,

$$\alpha = \lim_{z \to 1}[(z - 1)f(z)] = \lim_{z \to 1}\frac{z}{z + 1} = \frac{1}{1 + 1} = \frac{1}{2}$$

and

$$\oint_C \frac{z}{z^2 - 1} dz = 2\pi i \left(\frac{1}{2}\right) = \pi i$$

For the integral around the circle $|z| = 2$, we would have

$$\oint_{|z|=2} \frac{z}{z^2 - 1} dz = 2\pi i \left(\frac{1}{2} + \frac{-1}{-2}\right) = 2\pi i$$

Hitherto it has been assumed that the point of singularity of a function

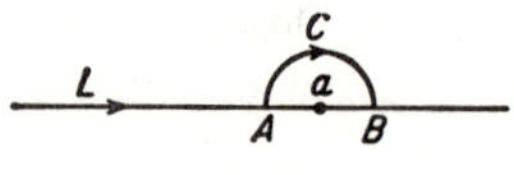

Fig. A-11

does not lie on the path of integration. Now let point a be a point on the line L (Fig. A-11), and consider the integral

$$\int_L \frac{f(z)}{z - a} dz$$

If $f(z) \neq 0$, the integrand becomes infinite as $z \to a$ and hence the integral is, in general, meaningless. To remedy this, we describe a circular arc with center at point a and of arbitrarily small radius r so as to cut the arc L at the two points A and B (Fig. A-11). Denoting the arc AB by C and the remaining length of line L by $L - C$, the above integral tends to a definite limit called the *Cauchy principal value* if there exists

$$\lim_{r \to 0} \int_{L-C} \frac{f(z)}{z - a} \, dz$$

In particular, if C is the upper half of the circle $|z - a| = r$ (Fig. A-12)

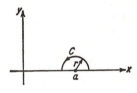

Fig. A-12

and a is on the real axis

$$\lim_{r \to 0} \oint_C f(z) \, dz = \pi i \, [\text{residue of } f(z) \text{ at } a] = \pi i \lim_{z \to a} [(z - a)f(z)] \quad (31)$$

Example 11. Evaluate the integral of $f(z) = z/[(1 - z)^2 \sqrt{z^2 - b^2}]$ around a small semicircle described in a clockwise sense with its center at $z = 1$.

From Eq. (31) we have directly

$$\oint \frac{z \, dz}{(1 - z^2) \sqrt{z^2 - b^2}} = -\pi i \lim_{z \to 1} \frac{z}{(1 + z) \sqrt{z^2 - b^2}} = -\frac{\pi i}{2 \sqrt{1 - b^2}}$$

RECOMMENDED EXERCISES

1. Evaluate the integral in Example 7 if C is the circle $|z| = \frac{1}{2}$; $|z - 1| = 3$.

2. Expand the function $1/(1 + z^2)$ in Maclaurin's series and determine the region of convergence.

3. Obtain the first four terms of Maclaurin's series for $\tan z$ and determine the radius of convergence for the series.

4. Expand the function z^2 in a power series in the neighborhood of $z = 1$.

5. Verify the results given in Eq. (28).

6. Evaluate the following integrals:

(a) $\displaystyle \oint_{|z|=2} \frac{z \, dz}{z^3 + 1}$

(b) $\displaystyle \oint_{|z|=2} \frac{ze^z \, dz}{z^2 - 1}$

(c) $\displaystyle \oint_C \frac{z \, dz}{z^2 + 2z + 5}$, for $C = (|z| = 3)$ and $C = (|z| = 1)$

(d) $\displaystyle \oint_{z=1} \frac{dz}{1 - z^2}$

Appendix B

B-1. Elliptic Integrals

The existence of the integral of a continuous function is assured; however, it does not necessarily follow that the integral can be expressed by elementary functions alone. In groundwater problems, as a consequence of the Schwarz-Christoffel transformation (Sec. 4-6), we often encounter integrals of the form

$$\int R[x, \sqrt{P(x)}] \, dx \qquad (1)$$

where $R(x)$ is a rational function of x, and $P(x)$ is a polynomial in x. In particular, if $P(x)$ is of third or fourth degree in x, and the integral cannot be expressed by elementary functions alone, Eq. (1) is called an *elliptic integral*. In the event that this integral can be expressed by elementary functions, for example

$$\int \frac{x \, dx}{\sqrt{x^4 + 1}} = \frac{1}{2} \ln (x^2 + \sqrt{x^4 + 1}) + C$$

the integral is called *pseudo-elliptic*.

There exists among engineers a general antipathy for calculations involving elliptic integrals. In large measure this is due to a lack of familiarity. Indeed, recognizing that all elliptic integrals can be reduced to three canonical forms, one needs only to consult the proper tables. In this connection the tables of Jahnke and Emde [65], Milne-Thomson [90], Nelson-Skornyakov [102], Selfridge and Maxfield [130], and Henderson [58] are highly recommended.

For further study of the theory of elliptic functions the works of Hancock [49, 50] and of Byrd and Friedman [12] are of great value.

B-2. Elliptic Integrals of the First Kind

The canonical forms of the elliptic integrals were first introduced by Legendre ("Traite des fonctions elliptiques," Paris) in 1825, and constitute the foundation of the theory of elliptic integrals and their associated elliptic functions.

The elliptic integral of the first kind in canonical form is

$$\int_0^x \frac{dx}{\sqrt{(1 - x^2)(1 - m^2x^2)}} \qquad (2)*$$

* In our work we will use m rather than the more common k to prevent confusion with k, the coefficient of permeability.

where m is called the *modulus* of the integral. Substituting $x = \sin \phi$ into Eq. (2), we have

$$F(m,\phi) = \int_0^\phi \frac{d\phi}{\sqrt{1 - m^2 \sin^2 \phi}} \qquad (3)^*$$

where $\phi = \sin^{-1} x$ is called the *amplitude* of the integral. $F(m,\phi)$ is the Legendre notation and is the usual tabular designation. In many tables the modulus is given as

$$m = \sin \theta \qquad (4)$$

where θ is called the *modular angle*. When used as such, Eq. (3) becomes

$$F(m,\phi) = \int_0^\phi \frac{d\phi}{\sqrt{1 - \sin^2 \theta \sin^2 \phi}} \qquad (5)$$

A particularly useful form of Eq. (3) is the *complete elliptic integral of the first kind*, defined as

$$K = F\left(m, \frac{\pi}{2}\right) = \int_0^{\pi/2} \frac{d\phi}{\sqrt{1 - m^2 \sin^2 \phi}} \qquad (6)$$

Example 1. Compute the value of the integral

$$\int_0^{0.5} \frac{dx}{\sqrt{(1 - x^2)(1 - 0.25x^2)}}$$

Making the change of variable $x = \sin \phi$, we obtain $\phi = 30°$. From the trigonometric form of $m^2 = 0.25$, we find $\theta = 30°$. Hence from, say, Jahnke and Emde (Table V) we have $F(30°,30°) = 0.5294$.

Example 2. Compute the value of the integral

$$\int_0^1 \frac{dx}{\sqrt{(1 - x^2)(1 - 0.25x^2)}}$$

Making the change of variable $x = \sin \phi$, we find $\phi = \pi/2$, and hence the required integral is the complete elliptic integral of the first kind of modulus $m = 0.5$. From Table B-1, we have for the integral $K = 1.686$.

Let us now consider two important special cases of elliptic integrals of the first kind:

1. *Modulus Greater than Unity, $m > 1$.* Taking the standard form of Eq. (2) as

$$\int_0^x \frac{dx}{\sqrt{(1 - x^2)(1 - m^2 x^2)}}$$

---

* Implied in the notation $F(m,\phi)$ is $F(m, \sin^{-1} x)$.

Table B-1. Complete Elliptic Integrals of the First Kind*

m^2	K	K'	$\dfrac{K}{K'}$	$\dfrac{K'}{K}$	m'^2	m^2	K	K'	$\dfrac{K}{K'}$	$\dfrac{K'}{K}$	m'^2
0.000	1.571	∞	0.000	∞	1.000	0.21	1.665	2.235	0.745	1.34	0.79
0.001	1.571	4.841	0.325	3.08	0.999	0.22	1.670	2.214	0.754	1.33	0.78
0.002	1.572	4.495	0.349	2.86	0.998	0.23	1.675	2.194	0.763	1.31	0.77
0.003	1.572	4.293	0.366	2.73	0.997	0.24	1.680	2.175	0.773	1.29	0.76
0.004	1.572	4.150	0.379	2.64	0.996	0.25	1.686	2.157	0.782	1.28	0.75
0.005	1.573	4.039	0.389	2.57	0.995	0.26	1.691	2.139	0.791	1.26	0.74
0.006	1.573	3.949	0.398	2.51	0.994	0.27	1.697	2.122	0.800	1.25	0.73
0.007	1.574	3.872	0.406	2.46	0.993	0.28	1.702	2.106	0.808	1.24	0.72
0.008	1.574	3.806	0.413	2.42	0.992	0.29	1.708	2.090	0.817	1.22	0.71
0.009	1.574	3.748	0.420	2.38	0.991	0.30	1.714	2.075	0.826	1.21	0.70
0.01	1.575	3.696	0.426	2.35	0.99	0.31	1.720	2.061	0.834	1.20	0.69
0.02	1.579	3.354	0.471	2.12	0.98	0.32	1.726	2.047	0.843	1.19	0.68
0.03	1.583	3.156	0.502	1.99	0.97	0.33	1.732	2.033	0.852	1.17	0.67
0.04	1.587	3.016	0.526	1.90	0.96	0.34	1.738	2.020	0.860	1.16	0.66
0.05	1.591	2.908	0.547	1.83	0.95	0.35	1.744	2.008	0.869	1.15	0.65
0.06	1.595	2.821	0.565	1.77	0.94	0.36	1.751	1.995	0.877	1.14	0.64
0.07	1.599	2.747	0.582	1.72	0.93	0.37	1.757	1.983	0.886	1.13	0.63
0.08	1.604	2.684	0.598	1.67	0.92	0.38	1.764	1.972	0.895	1.12	0.62
0.09	1.608	2.628	0.612	1.63	0.91	0.39	1.771	1.961	0.903	1.11	0.61
0.10	1.612	2.578	0.625	1.60	0.90	0.40	1.778	1.950	0.911	1.10	0.60
0.11	1.617	2.533	0.638	1.57	0.89	0.41	1.785	1.939	0.920	1.09	0.59
0.12	1.621	2.493	0.650	1.54	0.88	0.42	1.792	1.929	0.929	1.08	0.58
0.13	1.626	2.455	0.662	1.51	0.87	0.43	1.799	1.918	0.938	1.07	0.57
0.14	1.631	2.421	0.674	1.48	0.86	0.44	1.806	1.909	0.946	1.06	0.56
0.15	1.635	2.389	0.684	1.46	0.85	0.45	1.814	1.899	0.955	1.05	0.55
0.16	1.640	2.359	0.695	1.44	0.84	0.46	1.822	1.890	0.964	1.04	0.54
0.17	1.645	2.331	0.706	1.42	0.83	0.47	1.829	1.880	0.973	1.03	0.53
0.18	1.650	2.305	0.716	1.40	0.82	0.48	1.837	1.871	0.982	1.02	0.52
0.19	1.655	2.281	0.726	1.38	0.81	0.49	1.846	1.863	0.991	1.01	0.51
0.20	1.660	2.257	0.735	1.36	0.80	0.50	1.854	1.854	1.000	1.00	0.50
m'^2	K'	K	$\dfrac{K'}{K}$	$\dfrac{K}{K'}$	m^2	m'^2	K'	K	$\dfrac{K'}{K}$	$\dfrac{K}{K'}$	m^2

* From V. I. Aravin, and S. Numerov, "Seepage Computations for Hydraulic Structures," Stpoitel'stvu i Arkhitekture, Moscow, 1955.

and making the substitutions $x = y/m$ and $y = \sin \psi$,

$$\frac{1}{m} \int_0^{y/m} \frac{dy}{\sqrt{(1 - y^2/m^2)(1 - y^2)}}$$

we obtain the relation

$$F(m,\phi) = \frac{1}{m} F\left(\frac{1}{m}, \psi\right) \tag{7}$$

where $\sin \psi = m \sin \phi$.

2. *Amplitude Equal to* $1/m(m < 1)$. Separating the integral into the sum of two integrals, we have

$$\int_0^{1/m} \frac{dx}{\sqrt{(1 - x^2)(1 - m^2x^2)}} = \int_0^1 \frac{dx}{\sqrt{(1 - x^2)(1 - m^2x^2)}}$$
$$+ \int_1^{1/m} \frac{dx}{\sqrt{(1 - x^2)(1 - m^2x^2)}}$$
$$= K + \int_1^{1/m} \frac{dx}{\sqrt{(1 - x^2)(1 - m^2x^2)}} \tag{8a}$$

Now, introducing into the second term of Eq. (8a)

$$x = \frac{1}{\sqrt{1 - z^2m'^2}} \tag{8b}$$

where m', defined by the equation $m'^2 = 1 - m^2$, is called the *complementary modulus* or *comodulus*, we find

$$\int_0^{1/m} \frac{dx}{\sqrt{(1 - x^2)(1 - m^2x^2)}} = K + iK' \tag{8c}$$

where K' is the complete elliptic integral of the first kind with the complementary modulus m'. That is, $K'(m) = K(m')$ (see Table B-1).

RECOMMENDED EXERCISES

1. Evaluate the following integrals:

(a) $\int_0^1 \frac{dx}{\sqrt{(1 - x^2)(1 - 0.36x^2)}}$ (b) $\int_0^1 \frac{dx}{\sqrt{(1 - x^2)}}$

(c) $\int_0^{0.25} \frac{dx}{\sqrt{(1 - x^2)(1 - 4x^2)}}$ (d) $\int_0^2 \frac{dx}{\sqrt{(4 - x^2)(1 - x^2)}}$

2. Demonstrate each of the following:

(a) $K(0) = K'(1) = \pi/2$ (b) $F(1,\phi) = \ln(\tan \phi + \sec \phi)$
(c) $K' = K$ when $m = \sqrt{2}/2$ (d) $K' = \sqrt{2} K$ when $m = \sqrt{2} - 1$
(e) $F(m, \infty) = iK'$

B-3. Elliptic Functions; Jacobian Functions

Considering the function

$$u = \int_0^x \frac{dx}{\sqrt{1 - x^2}} = \sin^{-1} x \tag{9a}$$

it is evident that $x = \sin u$ is a more convenient form of expression, particularly when u is complex. Now noting that Eq. (9a) is but a special form of the elliptic integral of the first kind with $m = 0$,

$$u = \int_0^x \frac{dx}{\sqrt{(1 - x^2)(1 - m^2 x^2)}} = \int_0^\phi \frac{d\phi}{\sqrt{1 - m^2 \sin^2 \phi}} \qquad (9b)$$

where $x = \sin \phi$, let us apply a similar line of reasoning to the general case.

Recalling that ϕ was defined as the amplitude of the integral, written

$$\phi = \text{am } u \qquad (9c)$$

we define

$$x = \sin \phi = \sin (\text{am } u) = \text{sn } u \qquad (9d)$$

Its associated functions are

$$\begin{aligned}
x = \sin \phi &= \text{sn } u \\
\sqrt{1 - x^2} = \cos \phi &= \text{cn } u \\
\sqrt{1 - m^2 x^2} = \Delta \phi &= \text{dn } u
\end{aligned} \qquad (9e)*$$

Equations (9e) are called the *Jacobian elliptic functions* and play a vital role in the solution of groundwater problems. From a pragmatic point of view, as in the case of the inverse of the sine in Eq. (9a), the notation $x = \text{sn } u$ can be thought of as asking the question, "To what upper limit must the elliptic integral of the first kind of modulus m be raised $[x = ?$ in Eq. (2)] to yield u?" For example, recalling the definition of the complete elliptic integral of the first kind and taking $K = u$, we find $\phi = \text{am } K = \pi/2$ or $x = 1$, and $\text{sn } K = 1$. Thus

$$\text{sn } K = 1 \qquad \text{cn } K = 0 \qquad \text{dn } K = m' \qquad (9f)$$

An excellent tabulation of the elliptic functions is given by Milne-Thomson [90].

The dependence of the modulus of the elliptic functions is generally implied; however, when special notice of the modulus is required, they can be written as

$$\text{sn } (u,m) \qquad \text{cn } (u,m) \qquad \text{dn } (u,m) \qquad (9g)†$$

From the definitions of Eqs. (9e) it follows at once that

$$\begin{aligned}
\text{sn}^2 u + \text{cn}^2 u &= 1 \\
\text{dn}^2 u + m^2 \text{sn}^2 u &= 1 \\
\text{dn}^2 u - m^2 \text{cn}^2 u &= m'^2
\end{aligned} \qquad (10a)$$

In addition to these quasi-trigonometric relationships, we note that if

---

* These are pronounced "ess en" of u, etc.

† In Milne-Thomson's tables, instead of the forms of Eqs. (9g) the functions are given as sn $(u|m)$, cn $(u|m)$, etc. The vertical line means that the modulus is squared, i.e., sn $(u,0.5) = $ sn $(u|0.25)$.

ϕ is changed to $-\phi$,

$$\text{sn}\,(-u) = -\,\text{sn}\,u, \text{cn}\,(-u) = \text{cn}\,u \quad \text{and} \quad \text{dn}\,(-u) = \text{dn}\,u$$

which demonstrate that sn (u) is an odd function whereas cn (u) and dn (u) are even. Also, when the modulus $m = 0$, sn $(u,0) = \sin u$, cn $(u,0) = \cos u$, and dn $(u,0) = 1$. In a similar manner we find that when the modulus $m = 1$, sn $(u,1) = \tanh u$ and

$$\text{cn}\,(u,1) = \text{dn}\,(u,1) = \frac{1}{\cosh u}$$

The addition formulas for the three elliptic functions are

$$\text{sn}\,(u \pm v) = \frac{\text{sn}\,u\,\text{cn}\,v\,\text{dn}\,v \pm \text{sn}\,v\,\text{cn}\,u\,\text{dn}\,u}{1 - m^2\,\text{sn}^2\,u\,\text{sn}^2\,v} \tag{10b}$$

$$\text{cn}\,(u \pm v) = \frac{\text{cn}\,u\,\text{cn}\,v \mp \text{sn}\,u\,\text{sn}\,v\,\text{dn}\,u\,\text{dn}\,v}{1 - m^2\,\text{sn}^2\,u\,\text{sn}^2\,v} \tag{10c}$$

$$\text{dn}\,(u \pm v) = \frac{\text{dn}\,u\,\text{dn}\,v \mp m^2\,\text{sn}\,u\,\text{sn}\,v\,\text{cn}\,u\,\text{cn}\,v}{1 - m^2\,\text{sn}^2\,u\,\text{sn}^2\,v} \tag{10d}$$

Substituting $v = \pm nK$ into the above, we have

$$\text{sn}\,(u \pm K) = \pm \frac{\text{cn}\,u}{\text{dn}\,u} \quad \text{cn}\,(u \pm K) = \mp \frac{m'\,\text{sn}\,u}{\text{dn}\,u} \quad \text{dn}\,(u \pm K) = \frac{m'}{\text{dn}\,u}$$

$$\text{sn}\,(u \pm 2K) = -\,\text{sn}\,u \quad \text{cn}\,(u \pm 2K) = -\,\text{cn}\,u \quad \text{dn}\,(u \pm 2K) = \text{dn}\,u$$

$$\text{sn}\,(u \pm 4K) = \text{sn}\,u \quad \text{cn}\,(u \pm 4K) = \text{cn}\,u \quad \text{dn}\,(u \pm 4K) = \text{dn}\,u \tag{10e}$$

Noting from Eq. $(8c)^*$ that sn $(K + iK') = 1/m$, we find upon substitution of $u \pm v = K + iK'$ into Eqs. (10b) to (10d) that the functions sn u, cn u, and dn u are doubly periodic. That is, in addition to having periods for real values of K, they possess periods for imaginary values of K. The periodicity of the elliptic functions can be summarized as follows: sn u has the periods $4K$ and $2iK'$, cn u has the periods $4K$ and $2K + 2iK'$, and dn u has the periods $2K$ and $4iK'$. Thus, just as trigonometric functions can be reduced to arguments between 0 and 90°, by means of Eqs. (10e) and similar expressions, the arguments of the elliptic functions can be obtained from tabulated values between 0 and K. The elliptic functions for complex arguments are

$$\text{sn}\,(iu,m) = i\,\frac{\text{sn}\,(u,m')}{\text{cn}\,(u,m')} \tag{11a}$$

$$\text{cn}\,(iu,m) = \frac{1}{\text{cn}\,(u,m')} \tag{11b}$$

$$\text{dn}\,(iu,m) = \frac{\text{dn}\,(u,m')}{\text{cn}\,(u,m')} \tag{11c}$$

* See also Exercise 1(d), this section.

Equations (11) demonstrate that the functions cn and dn are real and sn is imaginary for imaginary arguments. From consideration of Eqs. (10) and (11) it follows that*

$$\text{sn} \ (iu \pm K) = \pm \ \frac{\text{cn} \ iu}{\text{dn} \ iu} = \pm \ \frac{1}{\text{dn} \ (u,m')} \tag{12a}$$

$$\text{cn} \ (iu \pm K) = \mp \ \frac{im' \ \text{sn} \ (u,m')}{\text{dn} \ (u,m')} \tag{12b}$$

$$\text{dn} \ (iu \pm K) = \frac{m' \ \text{cn} \ (u,m')}{\text{dn} \ (u,m')} \tag{12c}$$

In many problems the modulus may be greater than 1 or imaginary. It can be shown that these can be treated as is $0 \leqq m \leqq 1$ by a simple change of parameters. Considering $m \geqq 0$ and making the substitutions

$$\mu^2 = \frac{|m^2|}{1 + |m^2|} \qquad \mu_1{}^2 = \frac{1}{1 + |m^2|} \qquad v = \frac{u}{\mu_1} \tag{13a}$$

we have, noting $0 \leqq \mu \leqq 1$,

$$\text{sn} \ (u,im) = \mu_1 \ \frac{\text{sn} \ (v,\mu)}{\text{dn} \ (v,\mu)} \tag{13b}$$

$$\text{cn} \ (u,im) = \frac{\text{cn} \ (v,\mu)}{\text{dn} \ (v,\mu)} \tag{13c}$$

$$\text{dn} \ (u,im) = \frac{1}{\text{dn} \ (v,\mu)} \tag{13d}$$

For $m > 1$, the elliptic functions can be reduced to arguments with moduli $1/m < 1$ by the following formulas:

$$\text{sn} \ (u,m) = \frac{1}{m} \ \text{sn} \left(um, \frac{1}{m} \right) \tag{13e}$$

$$\text{cn} \ (u,m) = \text{dn} \left(um, \frac{1}{m} \right) \tag{13f}$$

$$\text{dn} \ (u,m) = \text{cn} \left(um, \frac{1}{m} \right) \tag{13g}$$

For additional formulations see Refs. 12, 49, 50, 65, 90, 106.

Example 3.† Find sn (u,m), cn (u,m), and dn (u,m), if $u = 0.5044$ and $m^2 = 0.8$. Here the argument lies between the tabular values of $u = 0.50$ and $u = 0.51$, with $m^2 = 0.8$. Using linear interpolation, we have

$$\begin{array}{l} \text{sn} \ (0.50|0.8) = 0.46557 \\ \text{sn} \ (0.51|0.8) = 0.47358 \end{array} \qquad \Delta = 0.00801$$

---

* For a tabulation of the elliptic functions of complex arguments see Ref. 58.
† All tabular values of the elliptic functions are from Ref. 90.

hence, $0.46557 + 0.44 \times 0.00801 = 0.46909$, and sn $(0.5044|0.8) = 0.46909$. By a similar procedure or from cn $u = (1 - \text{sn}^2 u)^{\frac{1}{2}}$ we obtain cn $(0.5044|0.8) = 0.88313$. Likewise, from tabular values or dn $u = (1 - m^2 \text{sn}^2 u)^{\frac{1}{2}}$, we find dn $(0.5044|0.8) = 0.90771$.

Generally a linear interpolation (or extrapolation) from the closest tabular values will suffice. However, should greater precision be desired, a second-order approximation such as is given in Ref. 90 may be used.

Example 4.* Evaluate sn $(5.0|0.7)$.

Since the argument lies beyond the tabulated values (like sin 255°) we must first reduce it to the range of the tables. Recalling from Eqs. (10e) that sn $(u - 2K) = -\text{sn}(u)$, we obtain $5.0 - 2K = 5.0 - 4.15073 = 0.84927$, and hence we have sn $(5.0|0.7) = -\text{sn}(0.84927|0.7) = -0.70905$.

Example 5.* Evaluate cn $(0.78|0.82)$.

We first find

$$\begin{aligned} \text{cn}\,(0.78|0.80) &= 0.74868 \\ \text{cn}\,(0.78|0.90) &= 0.75317 \end{aligned} \qquad \Delta = 0.00449$$

Then, using linear interpolation, we obtain $0.74868 + 0.2 \times 0.00449 = 0.74958$, and cn $(0.78|0.82) = 0.74958$.

Example 6.* Evaluate dn $(\frac{1}{2},i)$.

Noting that the modulus is imaginary, from Eqs. (13) we find $\mu = \sqrt{2}/2$, $\mu_1 = \sqrt{2}/2$, and $v = \sqrt{2}/2$. Hence dn $(\frac{1}{2},i) = 1/\text{dn}\,(\sqrt{2}/2, \sqrt{2}/2) = 1.11662$.

RECOMMENDED EXERCISES

1. Verify the following:

(a)   sn $(2K) = 0$ cn $(2K) = -1$ dn $(2K) = 1$
      sn $(4K) = 0$ cn $(4K) = 1$ dn $(4K) = 1$

(b)   sn $(u \pm 2K', m') = -\text{sn}(u,m')$
      sn $(u \pm 4K', m') = \text{sn}(u,m')$

(c)   sn $2u = \dfrac{2\,\text{sn}\,u\,\text{cn}\,u\,\text{dn}\,u}{1 - m^2 \text{sn}^4 u}$ (d)   cn $(K + iK') = -\dfrac{im'}{m}$

      cn $2u = \dfrac{\text{cn}^2 u - \text{sn}^2 u\,\text{dn}^2 u}{1 - m^2 \text{sn}^4 u}$ dn $(K + iK') = 0$

      dn $2u = \dfrac{\text{dn}^2 u - m^2 \text{sn}^2 u\,\text{cn}^2 u}{1 - m^2 \text{sn}^4 u}$

2. Evaluate the following:

(a)   sn $(12|0.5)$ (b)   cn $(1.0|0.85)$
(c)   dn $(7.0|0.36)$ (d)   dn $(4.0, 0.16)$
(e)   sn $(0.84,2i)$ (f)   sn $(0.5i + 4.0|0.6)$
(g)   sn^{-1} $(0.75|0.7)$

B-4. Elliptic Integrals of the Second Kind

The elliptic integral of the second kind in the canonical form of Legendre is defined as

$$\int_0^x \sqrt{\frac{1 - m^2 x^2}{1 - x^2}}\, dx \qquad (14a)$$

* See previous footnote.

Table B-2. Partial Table of Complete Elliptic Integrals of Second Kind*

$\dfrac{E}{K'-E'}$	m^2	E	E'	m'^2	
0.00	0.00	1.5708	1.0000	1.00	
0.5847	0.01	1.5669	1.0160	0.99	128.87
0.6721	0.02	1.5629	1.0286	0.98	64.989
0.7368	0.03	1.5589	1.0399	0.97	43.636
0.7911	0.04	1.5550	1.0505	0.96	32.932
0.8393	0.05	1.5509	1.0605	0.95	26.492
0.8836	0.06	1.5470	1.0700	0.94	22.187
0.9250	0.07	1.5429	1.0791	0.93	19.104
0.9644	0.08	1.5389	1.0880	0.92	16.785
1.0023	0.09	1.5348	1.0965	0.91	14.976
1.0387	0.10	1.5307	1.1050	0.90	13.525
1.0748	0.11	1.5267	1.1128	0.89	12.334
1.1098	0.12	1.5225	1.1207	0.88	11.339
1.1443	0.13	1.5184	1.1284	0.87	10.495
1.1785	0.14	1.5143	1.1360	0.86	9.7685
1.2123	0.15	1.5101	1.1434	0.85	9.1373
1.2460	0.16	1.5059	1.1506	0.84	8.5832
1.2796	0.17	1.5017	1.1578	0.83	8.0928
1.3131	0.18	1.4975	1.1648	0.82	7.6555
1.3467	0.19	1.4933	1.1717	0.81	7.2629
1.3804	0.20	1.4890	1.1785	0.80	6.9084
1.4142	0.21	1.4848	1.1852	0.79	6.5865
1.4483	0.22	1.4805	1.1918	0.78	6.2929
1.4826	0.23	1.4761	1.1983	0.77	5.9938
1.5172	0.24	1.4718	1.2047	0.76	5.7765
1.5521	0.25	1.4675	1.2111	0.75	5.5480
1.5875	0.26	1.4631	1.2173	0.74	5.3363
1.6232	0.27	1.4587	1.2235	0.73	5.1395
1.6595	0.28	1.4543	1.2300	0.72	4.9561
1.6961	0.29	1.4498	1.2357	0.71	4.7847
1.7337	0.30	1.4454	1.2417	0.70	4.6413
1.7716	0.31	1.4409	1.2476	0.69	4.4731
1.8103	0.32	1.4364	1.2534	0.68	4.3311
1.8497	0.33	1.4318	1.2593	0.67	4.1971
1.8898	0.34	1.4273	1.2650	0.66	4.0704
1.9307	0.35	1.4227	1.2707	0.65	3.9503
1.9724	0.36	1.4181	1.2763	0.64	3.8367
2.0151	0.37	1.4134	1.2819	0.63	3.7285
2.0587	0.38	1.4088	1.2875	0.62	3.6256
2.1034	0.39	1.4041	1.2930	0.61	3.5275
2.1491	0.40	1.3994	1.2984	0.60	3.4338
2.1960	0.41	1.3946	1.3038	0.59	3.3443
2.2442	0.42	1.3899	1.3092	0.58	3.2586
2.2936	0.43	1.3851	1.3145	0.57	3.1764
2.3443	0.44	1.3802	1.3198	0.56	3.0976
2.3966	0.45	1.3754	1.3250	0.55	3.0218
2.4504	0.46	1.3705	1.3302	0.54	2.9490
2.5058	0.47	1.3656	1.3354	0.53	2.8788
2.5629	0.48	1.3606	1.3405	0.52	2.8112
2.6219	0.49	1.3557	1.3456	0.51	2.7459
2.6829	0.50	1.3506	1.3506	0.50	2.6829
m'^2	E'	E	m^2	$\dfrac{E}{K'-E'}$	

* Adapted from F. B. Nelson-Skornyakov, "Seepage in Homogeneous Media," Gosudarctvennoe Izd. Sovetskaya Nauka, Moscow, 1949.

where m is the modulus. Substituting $x = \sin\phi$ into Eq. (14a), we have

$$E(m,\phi) = \int_0^\phi \sqrt{1 - m^2 \sin^2\phi}\, d\phi \qquad (14b)*$$

where ϕ is the amplitude and $E(m,\phi)$ is Legendre's notation for the elliptic integrals of the second kind. As in the case of elliptic integrals of the first kind, many tables use, for the modulus, $m = \sin\alpha$ (cf. Ref. 65); that is,

$$E(\alpha,\phi) = \int_0^\phi \sqrt{1 - \sin^2\alpha \sin^2\phi}\, d\phi \qquad (14c)$$

* As in the case of the first kind, implied in the notation $E(m,\phi)$ is $E(m, \sin^{-1} x)$.

The complete elliptic integral of the second kind $\mathbf{E}$ is defined as

$$\mathbf{E} = E\left(m, \frac{\pi}{2}\right) = \int_0^{\pi/2} \sqrt{1 - m^2 \sin^2 \phi}\, d\phi \qquad (15)$$

A short table of the complete elliptic integral of the second kind is given in Table B-2.

Example 7. Evaluate the integral

$$\int_0^{0.52} \sqrt{\frac{1 - 0.75x^2}{1 - x^2}}\, dx$$

Here, the amplitude is $\phi = \sin^{-1} 0.52 = 31°20'$ and the modulus is $m = \sqrt{3}/2 = \sin \alpha;\ \alpha = 60°$. Hence from Table V, Ref. 65 we have (with linear interpolation)

$$E(60°,31°20') = 0.5218 + {}^{2}\!\%_0 \times 0.0155 = 0.5270$$

Example 8. Evaluate the integral

$$\int_0^1 \sqrt{\frac{1 - 0.25x^2}{1 - x^2}}\, dx$$

Recognizing this integral as the complete elliptic integral of the second kind of modulus $m = 0.5$, we find directly from Table B-2,

$$\mathbf{E} = 1.4675$$

As in the case of the integrals of the first kind, we have the two important special cases:

$$E\left(\frac{1}{m}, \phi\right) = \frac{E(m,\theta) - m'^2 F(m,\theta)}{m} \qquad (16a)$$

where $\sin \phi = m \sin \theta$, and

$$E(m,\phi) = E(m,1/m) = \mathbf{E} + i(K' - \mathbf{E}') \qquad (16b)$$

where $\mathbf{E}'$ is the complete elliptic integral of the second kind of modulus m'.

In addition to their use in evaluating integrals such as Eqs. (14), elliptic integrals of the second kind are of value because of their connection with the integrals of the elliptic functions. To illustrate: recalling that $(1 - m^2 \sin^2 \phi)^{1/2} = \operatorname{dn} u$ where $\sin \phi = \operatorname{sn} u$, and

$$d\phi = \sqrt{1 - m^2 \sin^2 \phi}\, du$$

from Eq. (9b), and substituting into Eq. (14b), we obtain

$$E(m,\phi) = E(u) = \int_0^u \operatorname{dn}^2 u\, du \qquad (17a)$$

$E(u)$ is Jacobi's notation for the elliptic integral of the second kind as defined in Eq. (17a). As a consequence of this expression, the value of

$E(u)$ when $u = \pm K$ is

$$\pm \mathbf{E} = E(\pm K) = \int_0^{\pm K} \mathrm{dn}^2\, u\, du \tag{17b}$$

and when $u = K + iK'$, from Eq. (16b),

$$E(\pm K + iK') = \pm \mathbf{E} + i(K' - \mathbf{E}') \tag{17c}$$

Also, from definition (17a), we obtain

$$\int_0^u \mathrm{sn}^2\, u\, du = \int_0^u \frac{1 - \mathrm{dn}^2\, u\, du}{m^2} = \frac{1}{m^2}[F(m,\phi) - E(m,\phi)] \tag{17d}$$

The addition theorem for the elliptic integral of the second kind is

$$E(u + v) = E(u) + E(v) - m^2\, \mathrm{sn}\, u\, \mathrm{sn}\, v\, \mathrm{sn}\,(u + v) \tag{18a}$$

Putting $v = K$, we have

$$E(u + K) = E(u) + \mathbf{E} - \frac{m^2\, \mathrm{sn}\, u\, \mathrm{cn}\, u}{\mathrm{dn}\, u} \tag{18b}$$

Now letting $u = K$ in Eq. (18b), we obtain $E(2K) = 2\mathbf{E}$, whence

$$E(u + 2K) = E(u) + 2\mathbf{E}$$

and in general,

$$E(2kK \pm u) = 2k\mathbf{E} \pm E(u) \qquad k = \text{integer} \tag{18c}$$

Equation (18c) provides the means of extending the range of tabulated values of the elliptic integrals of the second kind. Using Legendre's notation, Eq. (18c) can be written as

$$E(m, k\pi \pm \phi) = 2k\mathbf{E} \pm E(m,\phi) \tag{18d}$$

For complex arguments

$$E(m,iv) = i\left[v - E'(v) + \frac{\mathrm{sn}\, v'\, \mathrm{dn}\, v'}{\mathrm{cn}\, v'}\right] \tag{19a}$$

where the prime denotes the modulus m'; i.e., $\mathrm{sn}\, v' = \mathrm{sn}\,(v,m')$. Substituting iv for v in Eq. (18a), one may obtain many useful forms, such as

$$E(m, K + iv) = \mathbf{E} + i\left[v - E'(v) + \frac{m'^2\, \mathrm{sn}\, v'\, \mathrm{cn}\, v'}{\mathrm{dn}\, v'}\right] \tag{19b}$$

$$E(m, u + iK') = E(u) + i(K' - \mathbf{E}') + \frac{\mathrm{cn}\, u\, \mathrm{dn}\, u}{\mathrm{sn}\, u} \tag{19c}$$

$$E(m, u + K + iK') = E(u) + \mathbf{E} + i(K' - \mathbf{E}') - \frac{\mathrm{sn}\, u\, \mathrm{dn}\, u}{\mathrm{cn}\, u} \tag{19d}$$

Finally we have Legendre's celebrated formula:

$$EK' + E'K - KK' = \frac{\pi}{2} \tag{20}$$

For additional relationships and tabulations, see Refs. 6, 10, 12, 49, 50, 58, 65, 106.

RECOMMENDED EXERCISES

1. Evaluate the following:

(a) $\int_0^{0.75} \sqrt{\frac{1 - 0.5x^2}{1 - x^2}}\, dx$ (b) $\int_0^{\sqrt{2}} \sqrt{\frac{1 - 0.5x^2}{1 - x^2}}\, dx$

(c) $E(m,u) = E(0.8,5)$

2. Verify Legendre's formula for an arbitrary value of m.

B-5. Jacobian Zeta Function

Jacobi defined the zeta function $Z(u)$ as

$$Z(u) = Z(u,m) = E(u) - \frac{E}{K}\, u \tag{21a}$$

It is immediately apparent from this definition that

$$Z(0) = 0 \qquad Z(K) = 0 \qquad Z(-u) = -Z(u) \tag{21b}$$

The addition formula

$$Z(u \pm v) = Z(u) \pm Z(v) \mp m^2 \operatorname{sn} u \operatorname{sn} v \operatorname{sn}(u \pm v) \tag{22}$$

is the same in form as that for the elliptic integral of the second kind [Eq. (18a)]. In particular, putting $v = K$ and $v = 2K$ in this expression, we have

$$Z(u + K) = Z(u) - m^2 \frac{\operatorname{sn} u \operatorname{cn} u}{\operatorname{dn} u} \tag{23a}$$

$$Z(u + 2K) = Z(u) \tag{23b}$$

Equation (23b) shows that the zeta function is simply periodic, with period $2K$.

Jacobi's *imaginary transformation*

$$Z(iu,m) = i\left[\frac{\operatorname{dn}(u,m')\operatorname{sn}(u,m')}{\operatorname{cn}(u,m')} - Z(u,m') - \frac{\pi u}{2KK'} \right] \tag{24}$$

together with the above relationships, permits the evaluation of the zeta functions for complex arguments.

An excellent table of zeta functions is given by Milne-Thomson [90]. For other important relationships and a tabulation of $KZ(u)$, see Byrd and Friedman [12].

RECOMMENDED EXERCISES

1. Obtain the graph of $Z(u)$ for $m = 0.8$, $0 \leq Z(u) \leq 4K$.

2. Verify:

    (a) $Z(u,1) = \tanh u$ (b) $Z(u + 2iK') = Z(u) - \dfrac{i\pi}{K}$

B-6. Elliptic Integrals of the Third Kind

The elliptic integral of the third kind in the canonical form of Legendre is

$$\Pi(m,n,x) = \int_0^x \frac{dx}{(1 + nx^2) \sqrt{(1 - x^2)(1 - m^2 x^2)}} \tag{25a}$$

or, as for the elliptic integrals of the first and second kind, substituting $x = \sin \phi$, we obtain

$$\Pi(m,n,\phi) = \int_0^\phi \frac{d\phi}{(1 + n \sin^2 \phi) \sqrt{1 - m^2 \sin^2 \phi}} \tag{25b}*$$

where, in addition to the modulus m and amplitude ϕ, the elliptic integral of the third kind is also dependent on the parameter n.

Taking $x = 1$ in Eq. (25a) or $\phi = \pi/2$ in Eq. (25b), we have the *complete elliptic integral* of the third kind,

$$\Pi(m, \pm n, 1) = \Pi_0(m, \pm n) = \int_0^1 \frac{dx}{(1 \pm nx^2) \sqrt{(1 - x^2)(1 - m^2 x^2)}} \tag{26a}$$

$$\Pi\left(m, \pm n, \frac{\pi}{2}\right) = \Pi_0(m, \pm n) = \int_0^{\pi/2} \frac{d\phi}{(1 \pm n \sin^2 \phi) \sqrt{1 - m^2 \sin^2 \phi}} \tag{26b}$$

Much of the difficulty of evaluating the elliptic integral of the third kind was removed in 1958 by the publication of "A Table of the Incomplete Elliptic Integral of the Third Kind" by Selfridge and Maxfield [130].[†] Prior to this tabulation, the integrals were calculated (often with considerable effort) from theta and zeta functions (cf. Refs. 12, 49, 102). Although the above tables are designed for the range $-1 < n < 1$, they

---

[*] In some literature n is taken as $-\alpha^2$ (cf. Refs. 12, 130). Note that $\Pi(m,n,\phi)$ implies $\Pi(m,n, \sin^{-1} x)$.

[†] The column headings in this table do not correspond with the authors' definition. To use the table, read $\alpha = -n$, $\kappa = m^2$, $\Theta = \phi$, as given in Eqs. (25) and (26). Thus for example, $\Pi(m,n,\phi) = \Pi(0.717, 0.2500, \pi/2)$ will be tabulated under $\Pi(\kappa,\alpha,\Theta) = \Pi(0.514, -0.2500, 1.5708) = 1.653$.

can be extended to include integrals with $|n| > 1$ by the formulas

$$\Pi(m, \pm n, \phi) + \Pi(m, \pm m^2/n, \phi) = F(m, \phi)$$
$$+ \frac{1}{\sqrt{(1 \pm n)(1 \pm m^2/n)}} \tan^{-1} \sqrt{\frac{(1 \pm n)(1 \pm m^2/n)}{1 - m^2 \sin^2 \phi}} \tan \phi \quad (27a)$$

$$\Pi_0(m, \pm n) + \Pi_0(m, \pm m^2/n) = K(m) + \frac{\pi}{2\sqrt{(1 \pm n)(1 \pm m^2/n)}} \quad (27b)$$

As in the case of the elliptic integrals of the first and second kinds, the integral with amplitude $x = 1/m$ $(m < 1)$ is of particular interest. Taking the standard integral as the sum of two integrals, we obtain

$$\int_0^{1/m} \frac{dx}{(1 + nx^2)\sqrt{(1 - x^2)(1 - m^2x^2)}}$$
$$= \Pi_0(m, n) + \int_1^{1/m} \frac{dx}{(1 + nx^2)\sqrt{(1 - x^2)(1 - m^2x^2)}}$$

Considering the second integral on the R.H.S. and substituting

$$t = \frac{1}{m'}\sqrt{1 - m^2x^2} \qquad m' = \sqrt{1 - m^2} \qquad n' = -\frac{nm'2}{m^2 + n} \quad (28a)$$

we obtain after some manipulation,

$$\int_1^{1/m} \frac{dx}{(1 + nx^2)\sqrt{(1 - x^2)(1 - m^2x^2)}}$$
$$= \frac{im^2}{n + m^2} \int_0^1 \frac{dt}{(1 + n't^2)\sqrt{(1 - t^2)(1 - m'^2t^2)}}$$

and hence

$$\int_0^{1/m} \frac{dx}{(1 + nx^2)\sqrt{(1 - x^2)(1 - m^2x^2)}}$$
$$= \Pi_0(m, n) + \frac{im^2}{n + m^2} \Pi_0(m', n') \quad (28b)$$

Another relation of considerable importance is the value of the elliptic integral of the third kind between the limits of $1/m$ and ∞ $(m < 1)$. From Eqs. (215.12) and (337.01) of Ref. 12, we find for this case

$$\int_{1/m}^\infty \frac{dx}{(1 + nx^2)\sqrt{(1 - x^2)(1 - m^2x^2)}} = -K(m) + \Pi_0\left(m, \frac{m^2}{n}\right) \quad (28c)$$

For additional tabulations of formulas for the elliptic integral of the third kind, the work of Byrd and Friedman [12] is unparalleled.

B-7. Gamma and Beta Functions

Two functions that are of particular value when considering special transformations (such as triangles) are the general factorial functions of Euler, called the *gamma* and *beta functions*.

The gamma function $\Gamma(a)$ is defined as

$$\Gamma(a) = \int_0^\infty x^{a-1} e^{-x}\, dx \qquad (29a)$$

or

$$\Gamma(a + 1) = \int_0^\infty x^a e^{-x}\, dx \qquad (29b)$$

The beta function is

$$\mathrm{B}(a,b) = \int_0^1 x^{a-1}(1 - x)^{b-1}\, dx \qquad a, b > 0 \qquad (30)$$

Tables of gamma functions are readily available in engineering literature (cf. page 14 of Ref. 65). Beta functions are generally determined from gamma functions by

$$\mathrm{B}(a,b) = \frac{\Gamma(a)\Gamma(b)}{\Gamma(a + b)} \qquad (31a)$$

Also, it can be shown that

$$\mathrm{B}(a,b) = \frac{b - 1}{a}\,\mathrm{B}(a + 1,\, b - 1) \qquad (31b)$$

One of the major uses of the gamma function is derived from its relationship to the familiar factorial,

$$n! = n(n - 1)(n - 2)\,\cdots\,(3)(2)(1)$$

Whereas the factorial is confined to positive integral values of n only, the gamma function extends the concept of a factorial to all values, positive and negative, integral and fractional, excepting negative integers. This correspondence is provided by the recurrence formula

$$\Gamma(a + 1) = a\Gamma(a) \qquad (32)$$

To illustrate, noting in Eq. (29a) that when $a = 1$,

$$\Gamma(1) = \int_0^\infty e^{-x}\, dx = 1$$

we obtain

$$\Gamma(1) = 1$$
$$\Gamma(2) = 1\Gamma(1) = 1!$$
$$\Gamma(3) = 2\Gamma(2) = 2!$$
$$\Gamma(4) = 3\Gamma(3) = 3!$$

and in general

$$\Gamma(n + 1) = n! \quad \text{or} \quad \Gamma(n) = (n - 1)! \tag{33}$$

Another useful relationship follows from Euler's integral,

$$\int_0^\infty e^{-x^2}\, dx = \frac{\sqrt{\pi}}{2}$$

Substituting in this $x = t^{1/2}$ and multiplying by 2, we find

$$\int_0^\infty e^{-t} t^{1/2-1}\, dt = \Gamma(1/2) = \sqrt{\pi}$$

$$\Gamma\left(\frac{3}{2}\right) = \frac{1}{2}\, \Gamma\left(\frac{1}{2}\right) = \frac{\sqrt{\pi}}{2}$$

$$\Gamma\left(\frac{5}{2}\right) = \frac{3}{2}\, \Gamma\left(\frac{3}{2}\right) = \frac{1 \cdot 3}{2^2}\, \sqrt{\pi}$$

$$\Gamma\left(\frac{7}{2}\right) = \frac{5}{2}\, \Gamma\left(\frac{5}{2}\right) = \frac{1 \cdot 3 \cdot 5}{2^3}\, \sqrt{\pi}$$

and in general

$$\Gamma\left(n + \frac{1}{2}\right) = \frac{1 \cdot 3 \cdot 5 \, \cdots \, (2n - 1)}{2^n}\, \sqrt{\pi} \tag{34}$$

On the basis of the above, we obtain very simply,

$$\Gamma\left(-\frac{1}{2}\right) = \frac{\Gamma(-1/2 + 1)}{-1/2} = \frac{\Gamma(1/2)}{-1/2} = -2\, \sqrt{\pi}$$

$$\Gamma\left(-\frac{3}{2}\right) = \frac{\Gamma(-3/2 + 1)}{-3/2} = \frac{\Gamma(-1/2)}{-3/2} = \frac{4}{3}\, \sqrt{\pi}$$

Two other valuable relationships are:

$$\Gamma(a)\Gamma(1 - a) = \frac{\pi}{\sin \pi a} \tag{35a}$$

$$B(p,q) = 2 \int_0^{\pi/2} \sin^{2p-1} z \, \cos^{2q-1} z \, dz \tag{35b}$$

RECOMMENDED EXERCISES

1. Evaluate:

(a) $\Gamma(0)$ (b) $\Gamma(1.63)$

(c) $\displaystyle\int_0^\infty \frac{x^a}{a^x}\, dx$ (Hint: $a^{-x} = e^{-x \ln a}$)   (d) $B(2,3)$

2. Verify Eq. (32).

3. Demonstrate that the gamma functions are undefined for negative integers.

References

The reference style common to Russian references has been retained to provide easier identification to Russian libraries. Russian names and sources have been transliterated according to the Library of Congress system.

DAN   Doklady Akademii Nauk
Izv. NIIG   Izvestiya, Nauchno-Issledovatel'skogo In-ta Gidrotekhniki
PMM   Prikladnaya Matematika i Mekhanika
ZAMM   Zeitschrift für angewandte Mathematik und Mechanik

1. ASCE: "Symposium on Rockfill Dams," Proc. Symposium Series no. 3, reprinted from *J. Power Div.*, September, 1958.
2. Aravin, V. I., and S. Numerov (Аравин, В. И., и С. Нумеров): "Seepage Computations for Hydraulic Structures," Gosstroĭizdat, Moscow, 1948.
3. Aravin, V. I., and S. Numerov: "Theory of Flow of Fluids and Gases in Incompressible Porous Media," Gostekhizdat, Moscow, 1953.
4. Aravin, V. I., and S. Numerov: "Seepage Computations for Hydraulic Structures," Stroitel'stvu i Arkhitekture, Leningrad, 1955.
5. Babbitt, H., and D. H. Caldwell: The Free Surface around, and Interference between, Gravity Wells, *Univ. Illinois Eng. Exp. Sta. Bull.* 374, 1948.
6. Erdélyi, A.: "Higher Transcendental Functions," vol. 2, McGraw-Hill Book Company, Inc., New York, 1953.
7. Bazanov, M. I. (Базанов, М. И.): Concerning the Seepage under Dams for Condition of Plane Flow, *Izv. NIIG*, vol. 28, 1940.
8. Bieberbach, L.: "Conformal Mapping," Chelsea Publishing Company, New York, 1953.
9. Bil'dyug, E. I. (Бильдюг, Е. И.): Computation for Seepage in an Earth Dam with Shield and Cutoff, *Ivz. NIIG*, vol. 28, 1940.
10. Bowman, F.: "Introduction to Elliptic Functions, with Applications," English University Press Ltd., London, 1953.
11. Breitenöder, M.: "Ebene Grundwasserströmungen mit freir Oberfläche," Springer-Verlag, Berlin, 1942.
12. Byrd, P. F., and M. D. Friedman: "Handbook of Elliptic Integrals for Engineers and Physicists," Springer-Verlag, Berlin, 1954.
13. Carman, P. C.: "Flow of Gases through Porous Media," Academic Press, Inc., New York, 1956.
14. Casagrande, A.: Seepage through Dams, in "Contributions to Soil Mechanics 1925–1940," Boston Society of Civil Engineers, Boston, 1940.
15. Casagrande, A., and W. L. Shannon: Base Course Drainage for Airport Pavements, *Trans. Am. Soc. Civil Engrs.*, vol. 117, 1952.
16. Casagrande, L.: "Naeherungsmethoden zur Bestimmung von Art und Menge der Sickerung durch geschuettete Daemme," Thesis, Technische Hochschule, Vienna, 1932. Translated by U.S. Corps of Engineers, Waterways Exp. Sta., Vicksburg, Miss.

17. Cayley, A.: "Elliptic Functions," Deighton, Bell and Co., London, 1876.
18. Charny, I. A. (Чарный, И. А.): "Basic Underground Hydraulics," Goctoptek-hizdat, Moscow, 1956.
19. Chow, V. T.: "Open-channel Hydraulics," McGraw-Hill Book Company, Inc., New York, 1959.
20. Churchill, R. V.: "Complex Variables and Applications," McGraw-Hill Book Company, Inc., New York, 1960.
21. Dachler, R.: Der Sickervorgang in Dammboschungen, *Die Wasserwirtschaft*, no. 4, 1933.
22. Dachler, R.: Ueber die Versickerung aus Kanälen, *Die Wasserwirtschaft*, no. 9, 1933.
23. Dachler, R.: Ueber den Strömungsvorgang bei Hangquellen, *Die Wasserwirtschaft*, no. 5, 1934.
24. Dachler, R.: "Grundwasserströmung," Springer-Verlag, Vienna, 1936.
25. Darcy, H.: "Les Fontaines publiques de la ville de Dijon," Paris, 1856.
26. Davison, B. B. (Девисон, Б. Б.): Some Questions of Simplifying Computations of Confined Flow of Ground Water, *Trudi Gos Gidrol. Inst.*, no. 5, 1937.
27. De La Marre, P. H.: New Methods for the Experimental Calculation of Flow in Porous Media, *La Houille Blanche*, vol. 8, 1953.
28. Dupuit, J.: "Etudes théoriques et pratiques sur le mouvement des eaux dans les canaux découverts et à travers les terrains perméables," Paris, 1863.
29. Dwight, H. B.: "Tables of Integrals and other Mathematical Data," rev. ed., The Macmillan Company, New York, 1947.
30. Englund, F.: Mathematical Discussion of Drainage Problems, *Trans. Danish Acad. Tech. Sci.*, no. 3, 1951.
31. Englund, F.: On the Laminar and Turbulent Flow of Ground Water through Homogeneous Sand, *Trans. Danish Acad. Tech. Sci.*, no. 3, 1953.
32. Fandeev, V. V. (Фандеев, В. В.): Determination of Forces for the Calculation of the Stability of Slopes in Earth Dams, *Gidrotekh. Stroitel.*, no. 1, 1947.
33. Fil'chakov, P. F. (Фильчаков, П. Ф.): "Mechanics of Hydraulic Computations for Dams," Thesis, Matem. In-t Akad. Nauk USSR, 1951.
34. Fil'chakov, P. F.: "Theory of Seepage under Hydraulic Structures," Akad. Nauk Ukran. SSR, Kiev, 1960.
35. Fletcher, A.: "Guide to Tables of Elliptic Functions, Mathematic Tables and Other Aids to Computation," vol. 3, National Research Council, Washington, D.C., 1948.
36. Forchheimer, P.: Über den freien Spiegel bei Grundwasserströmungen, *ZAMM*, no. 2, 1922.
37. Forchheimer, P.: "Hydraulik," Teubner Verlagsgesellschaft, mbh, Stuttgart, 1930.
38. Frocht, M. M.: "Photoelasticity," John Wiley & Sons, Inc., New York, 1948.
39. Garder, W., T. R. Collier, and D. Farr: Ground Water, *Utah Agr. Exp. Sta. Bull.*, no. 252, Logan, Utah, 1934.
40. Garmonov, I. V., and A. V. Lebedev (Гармонов, И. В. и А. В. Лебедев): "Basic Problems of the Dynamics of Underground Water," Gosudat. Izdatel. Geol. Liter, Moscow, 1952.
41. Gibbs, W. J.: "Conformal Transformations in Electrical Engineering," Chapman & Hall, Ltd., London, 1958.
42. Gilboy, G.: Hydraulic-Fill Dams, *Proc. Intern. Comm. Large Dams*, Stockholm, 1933.
43. Girinsky, N. K. (Гиринский, Н. К.): Fundamental Theory of the Flow of Ground

Water under Hydraulic Structures in the Presence of a Constant Pressure along the Lower Soil Surface, *Gidrotekh. Stroitel.*, no. 6, 1936.

44. Girinsky, N. K.: Determination of the Coefficient of Permeability, *Gosgeolizdat,* 1950.
45. Graton, L. C., and H. J. Fraser: Systematic Packing of Spheres with Particular Relation to Porosity and Permeability, and Experimental Study of the Porosity and Permeability of Clastic Sediments, *J. Geol.*, vol. 43, 1935.
46. Hall, H. P.: A Historical Review of Investigations of Seepage toward Wells, *J. Boston Soc. Civil Eng.*, vol. 41, no. 3, 1954.
47. Hall, H. P.: An Investigation of Steady Flow toward a Gravity Well, *La Houille Blanche*, January-February, 1955.
48. Hamel, G.: Ueber Grundwasserströmung, *ZAMM*, vol. 14, 1934.
49. Hancock, H.: "Elliptic Integrals," Dover Publications, New York, 1958.
50. Hancock, H.: "Theory of Elliptic Functions," Dover Publications, New York, 1958.
51. Hansen, V. E.: Unconfined Ground-water Flow to Multiple Wells, *Trans. Am. Soc. Civil Engrs.*, vol. 118, 1953.
52. Harr, M. E., and S. P. Brahma: Seepage into a Filter Considering Reverse Flow, *Symposium of Indian Natl. Soc. of Soil Mech. and Foundation Eng.*, New Delhi, 1961.
53. Harr, M. E., and R. C. Deen: On the Analysis of Seepage Problems, *J. Soil Mech. and Foundation Div.*, Am. Soc. Civil Engrs., October, 1961.
54. Harza, L. F.: Uplift and Seepage under Dams on Sand, *Trans. Am. Soc. Civil Engrs.*, vol. 100, 1935.
55. Hele-Shaw, H. S.: Experiments on the Nature of the Surface Resistance in Pipes and on Ships, *Trans. Inst. Naval Architects*, vol. 39, 1897.
56. Hele-Shaw, H. S.: Investigation of the Nature of Surface Resistance of Water and of Stream-Line Motion under Certain Experimental Conditions, *Trans. Inst. Naval Architects*, vol. 40, 1898.
57. Hele-Shaw, H. S.: Stream-Line Motion of a Viscous Film, *68th Meeting, British Association for the Advancement of Science*, 1899.
58. Henderson, F. M.: "Elliptic Functions with Complex Arguments," University of Michigan Press, Ann Arbor, Mich., 1960.
59. Hopf, L., and E. Trefftz: Grundwasserströmung in einem abfallenden Gelände mit Abfanggraben, *ZAMM*, vol. 1, 1921.
60. Hubbert, M. K.: The Theory of Ground-water Motion, *J. Geol.*, vol. 48, no. 8, 1940.
61. Istomina, V. S. (Истомина, В. С.): "Steady Seepage in Soils," Gos. Izd. Liter. Stroitel. i Arkhit., Moscow, 1957.
62. Iterson, F. K. Th. van: Eenige Theoretische Beschouwingen over kwel, *De Ingenieur*, 1916 and 1917.
63. Jacob, C. E.: Drawdown Test to Determine Effective Radius of Artesian Well, *Trans. Am. Soc. Civil Engrs.*, vol. 112, 1947.
64. Jaeger, C.: "Engineering Fluid Mechanics," Blackie & Son Ltd., Glasgow, 1956.
65. Jahnke, E., and F. Emde: "Tables of Functions," 4th ed., Dover Publications, New York, 1945.
66. Kamensky, G. N. (Каменский, Г. Н.): Flow of Groundwater in a Soil Mass between Rivers, *DAN USSR*, vol. 21, no. 5, 1938.
67. Kaplan, W.: "Functions of a Complex Variable," Addison-Wesley Publishing Company, Reading, Mass., 1953.
68. Kellogg, O. D.: "Foundations of Potential Theory," Frederick Ungar Publishing Co., New York, 1929.

69. Khosla, R. B. A. N., N. K. Bose, and E. McK. Taylor: "Design of Weirs on Permeable Foundations," Central Board of Irrigation, New Delhi, India, 1954.

70. Kristianovich, S. A., S. G. Mikhlin, and B. B. Davison (Христианович, С. А., С. Г. Михлин, и Б. Б. Девисон): "On Some Problems on the Mechanics of Continuous Media," Izd. Akad. Nauk SSSR, Moscow, 1938.

71. Kirkham, D.: Artificial Drainage of Land: Streamline Experiments, the Artesian Basin, III, *Trans. Am. Geophys. Union*, vol. 26, 1945.

72. Kirkham, D.: Seepage into Drain Tubes in Stratified Soil, *Trans. Am. Geophys. Union*, vol. 32, 1951.

73. Kober, H.: "Dictionary of Conformal Representations," Dover Publications, 1957.

74. Koppenfels, W. von: "Praxis der konformen abbildung," Springer-Verlag, Berlin, 1959.

75. Kozeny, J.: Ueber kapillare Leitung des Wassers im Boden, Wien. Akad. Wiss., 1927.

76. Kozeny, J.: Grundwasserbewegung bei freiem Spiegel, Fluss und Kanalversickerung, *Wasserkraft und Wasserwirtschaft*, no. 3, 1931.

77. Kozeny, J.: Theorie und Berechnung der Brunnen, *Wasserkraft und Wasserwirtschaft*, nos. 8–10, 1933.

78. Kozeny, J.: "Hydraulik," Springer-Verlag, Berlin, 1953.

79. Kozlov, V. S. (Козлов, В. С.): "Hydromechanical Computations for Weirs," Gosenergoizdat, Moscow-Leningrad, 1941.

80. Lamb, H.: "Hydrodynamics," 6th ed., Dover Publications, New York, 1945.

81. Lambe, T. W.: "Soil Testing for Engineers," John Wiley & Sons, Inc., New York, 1951.

82. Lane, E. W.: Security from Under-Seepage: Masonry Dams on Earth Foundations, *Trans. Am. Soc. Civil Engrs.*, vol. 100, 1935.

83. Lavrent'ev, M. A., and B. V. Shabat (Лаврентьев, М. А., и Б. В. Шабат): "Methods of the Theory of Functions of Complex Variables," Gostekhizdat, Moscow, 1951.

84. Leibenson, L. S. (Лейбензон, Л. С.): "Motion of Natural Liquids and Gases in a Porous Media," Gostekhizdat, Moscow, 1947.

85. Leliavsky, S.: "Percolation under Aprons of Irrigation Works," Cairo, 1929.

86. Leliavsky, S.: "Irrigation and Hydraulic Design," vol. 1, Chapman and Hall, Ltd., London, 1955.

87. Leonards, G. A.: "Foundation Engineering," McGraw-Hill Book Company, Inc., New York, 1962.

88. Luthin, J. N.: "Drainage of Agricultural Lands," American Society of Agronomy, Madison, Wis., 1957.

89. McNown, J. S., E. Hsu, and C. Yih: Applications of the Relaxation Technique in Fluid Mechanics, *Trans. Am. Soc. Civil Engrs.*, vol. 120, 1955.

90. Milne-Thomson, L. M.: "Jacobian Elliptic Function Tables," Dover Publications, New York, 1950.

91. Milne-Thomson, L. M.: "Theoretical Hydrodynamics," The Macmillan Company, New York, 1960.

92. Mikhaylov, G. K. (Михайлов, Г. К.): Concerning the Seepage in a Trapezoidal Dam on an Horizontal Impervious Base, *Gidrotekhnika i Melioratsiya*, no. 1, 1952.

93. Mkhitarian, A. M. (Мхитарян, А. М.): Seepage of Water through Earth Dams on Impervious Bases, *Izv. Akad. Nauk Arm. S.S.R.*, no. 5, 1947.

94. Mkhitarian, A. M.: Computation of Seepage through Earth Dams, *Inzhenernii Sbornik*, vol. 14, 1952.

95. Mkhitarian, A. M.: Computations for the Seepage through an Earth Dam with a Sheetpile and a Drain, *Inzhenernïi Sbornik*, vol. 15, 1953.
96. Mkhitarian, A. M.: Seepage through an Earth Dam on an Inclined Impervious Base, *Inzhenernïi Sbornik*, vol. 16, 1953.
97. Murphy, G.: "Similitude in Engineering," The Ronald Press Company, New York, 1950.
98. Murray, J. A.: Relaxation Methods Applied to Seepage Flow Problems in Earth Dams and Drainage Wells, *J. Inst. Engrs. (India)*, vol. 41, no. 4, 1960.
99. Muskat, M.: "The Flow of Homogeneous Fluids through Porous Media,' McGraw-Hill Book Company, New York, 1937; reprinted by J. W. Edwards, Publisher, Inc., Ann Arbor, 1946.
100. Muskhelishvili, N. I.: "Some Basic Problems of the Mathematical Theory of Elasticity," P. Noordhoff, N.V., Groningen, Netherlands, 1953.
101. Nehari, Z.: "Conformal Mapping," McGraw-Hill Book Company, Inc., New York, 1952.
102. Nelson-Skornyakov, F. B. (Нельсон-Скорняков, Ф. Б.): "Seepage in Homogeneous Media," Gosudarctvennoe Izd. Sovetskaya Nauka, Moscow, 1949.
103. Numerov, S. N. (Нумеров, С. Н.): On Seepage in Earth Dams with Drainage on Impervious Foundations, *Izv. NIIG*, vol. 25, 1939.
104. Numerov, S. N.: Solution of Problem of Seepage without Surface of Seepage and without Evaporation or Infiltration of Water from Free Surface, *PMM*, vol. 6, 1942.
105. Numerov, S. N.: Approximate Method of Computing the Seepage through an Earth Dam on a Porous Base, *Trudi Leningradskogo Politekhn. In-ta*, no. 4, 1947, no. 5, 1948.
106. Oberhettinger, F., and W. Magnus: "Anwendung der elliptischen Funktionen in Physik und Technik," Springer-Verlag, Berlin, 1949.
107. Panov, D. (Панов, Д.): "Numerical Methods of Integrating Partial Differential Equations," Gostekhizdat, Moscow, 1951.
108. Pavlovsky, N. N. (Павловский, Н. Н.): Seepage through Earth Dams, *Instit. Gidrotekhniki i Melioratsii*, Leningrad, 1931. Translated by U.S. Corps of Engineers.
109. Pavlovsky, N. N.: Motion of Water under Dams, *Trans. 1st Congr. on Large Dams*, Stockholm, vol. 4, 1933.
110. Pavlovsky, N. N.: "Collected Works," Akad. Nauk USSR, Leningrad, 1956.
111. Peirce, B. O.: "A Short Table of Integrals," 3d ed., Ginn and Company, Boston, 1929.
112. Pirverdian, A. M. (Пирвердян, А. М.): "Petroleum Underground Hydraulics," Aznefteizdat, Baku, 1956.
113. Polubarinova-Kochina, P. Ya. (Полубаринова-Кочина, П. Я.): On the Continuity of the Velocity Hodograph in Plane Steady Motion of Ground Water, *DAN*, vol. 24, no. 4, 1939.
114. Polubarinova-Kochina, P. Ya.: Concerning Seepage in Heterogeneous (Two-layered) Media, *Inzhenernïi Sbornik*, vol. 1, no. 2, 1941.
115. Polubarinova-Kochina, P. Ya.: On the Direct and Inverse Problems of the Hydraulics of an Oil Layer, *PMM*, vol. 7, no. 5, 1943.
116. Polubarinova-Kochina, P. Ya.: "Theory of the Motion of Ground Water," Gostekhizdat, Moscow, 1952.
117. Prandtl, L.: Über Flussigkeitsbewegung bei sehr kleiner Reibung, *Verhandl. III Intern. Math.-Kongr.*, Heidelberg, 1904.
118. Prášil, F.: "Technische Hydrodynamik," Springer-verlag, Berlin, 1913.
119. Privalov, I. I. (Привалов, И. И.): "Introduction to the Theory of Functions of Complex Variables," Fizmattiz, Moscow, 1960.

120. Rel'tov, B. F. (Рельтов, Б. Ф.): Study of Seepage in Spatial Problems by Electrical Analogues, *Izv. NIIG*, 1935. See also *Trans. 2d Congr. on Large Dams*, Washington, vol. 5, 1936.

121. Reynolds, O.: An Experimental Investigation of the Circumstances which Determine whether the Motion of Water shall be Direct or Sinuous and of the Law of Resistance in Parallel Channels, *Phil. Trans. Roy. Soc. London*, vol. 174, 1883.

122. Risenkampf, B. K. (Ризенкампф, Б. К.): Hydraulics of Soil Water, *Uchen. zap Saratovskogo goc. un-ta, Seriya mat. mekh.*, vol. 14, 1938, vol. 15, 1940.

123. Rouse, H., and S. Ince: "History of Hydraulics," Iowa Institute of Hydraulic Research, State University of Iowa, 1957.

124. Samsioe, A. F.: Einfluss von Rohrbrunnen auf die Bewegung des Grundwassers, *ZAMM*, vol. 11, 1931.

125. Schaffernak, F.: Über die Standsicherheit durchlaessiger geschuetteter Dämme, *Allgem. Bauzeitung*, 1917.

126. Schaffernak, F.: Erforschung der physikalischen Gesetze, nach welchen die Durchsicherung des Wassers durch eine Talsperre oder durch den Untergrund stattfindet, *Die Wasserwirtschaft*, no. 30, 1933.

127. Scheidegger, A. E.: "The Physics of Flow through Porous Media," The Macmillan Company, New York, 1957.

128. Schmid, W. E.: Water Movement in Soils under Pressure Potentials, *Highway Research Board Spec. Rept.* 40, Washington, 1958.

129. Segal, B. I. (Сегал, Б. И.): Approximate Calculation of some Hyperelliptic Integrals that Occur in the Design of Dams, *DAN*, vol. 35, no. 7, 1942.

130. Selfridge, R. G., and J. E. Maxfield: "A Table of the Incomplete Elliptic Integral of the Third Kind," Dover Publications, New York, 1958.

131. Selim, M. A.: Dams on Porous Media, *Trans. Am. Soc. Civil Engrs.*, vol. 112, 1947.

132. Shankin, P. A. (Шанкин, П. А.): "Computations for Seepage through Earth Dams," Rechizdat, Moscow, 1947.

133. Shchelkachev, V. N. (Щелкачев, В. Н.): "Principles of Underground Oil Hydraulics," Gostoptekhizdat, Moscow, 1945.

134. Shchelkachev, V. N., and G. B. Pykhachev (Г. Б. Пыхачев): "Interference Wells and the Theory of Stratified Water-pressure Systems," Aznefteizdat, Baku, 1939.

135. Silin-Bekchurin, A. I. (Силин-Бекчурин, А. И.): "Dynamics of Underground Water," Moscow University, Moscow, 1958.

136. Slichter, C. S.: Theoretical Investigation of the Motion of Ground Waters, *U.S. Geol. Survey, 19th Ann. Rept.*, part 2, 1899.

137. Smirnov, V. I. (Смирнов, В. И.): "Course in Higher Mathematic," Gosudar. Izd Fiz.-Matem. Liter., Moscow, 1958.

138. Sokolov, Yu. D. (Соколов, Ю. Д.): Concerning Seepage from a Canal of Trapezoidal Section, *DAN*, vol. 79, no. 5, 1951.

139. Southwell, R. V.: "Relaxation Methods in Theoretical Physics," Oxford University Press, New York, 1946.

140. Stevens, O.: Discussion of Ground Water and Seepage, *Proc. Intern. Conf. Soil Mech. and Foundation Eng.*, vol. 3, Cambridge, Mass., 1936.

141. Streeter, V. L.: "Fluid Dynamics," McGraw-Hill Book Company, Inc., 1948.

142. Taylor, D. W.: "Fundamentals of Soil Mechanics," John Wiley & Sons, Inc., New York, 1948.

143. Terzaghi, K. von: Der Grundbruch and Stauwerken und seine Verhütung, *Die Wasserkraft*, 1922.

144. Terzaghi, K.: "Theoretical Soil Mechanics," John Wiley & Sons, Inc., New York, 1943.

145. Terzaghi, K., and R. B. Peck: "Soil Mechanics and Engineering Practice," John Wiley & Sons, Inc., New York, 1948.

146. Todd, D. K.: "Investigation of Unsteady Flow in Porous Media by Means of a Hele-Shaw Viscous Fluid Model," Dissertation, University of California, Berkeley, 1953.

147. U.S. Bureau of Reclamation: Model Studies of Penstocks and Outlet Works, Boulder Canyon Project, Final Reports, Part VI, Bull. no. 2, Denver, 1938.

148. U.S. Bureau of Reclamation: "Design of Small Dams," U.S. Govt. Printing Office, Washington, 1960.

149. Vallentine, H. R.: "Applied Hydrodynamics," Butterworths Scientific Publications, London, 1959.

150. van Deemter, J. J.: Results of Mathematical Approach to Some Flow Problems Connected with Drainage and Irrigation, *Appl. Sci. Research*, 1949.

151. Vedernikov, V. V. (Ведерников, В. В.): Seepage from Channels, *Gosstroĭizdat*, 1934. See also Versickerungen aus Kanälen, *Wasserkraft und Wasserwirtschaft*, nos. 11, 12, and 13, 1934.

152. Vedernikov, V. V.: Theory of Seepage and Its Applications to Problems of Irrigation and Drainage, *Gosstroĭizdat*, 1939.

153. Vedernikov, V. V.: On Theory of Drainage, *DAN*, vol. 23, no. 4, 1939.

154. Vedernikov, V. V.: Seepage with the Presence of Drains or Permeable Layers, *DAN*, vol. 69, no. 5, 1949.

155. Verigin, N. N. (Веригин, Н. Н.): Seepage through the Foundation of Dams with Oblique Screens and Cutoffs, *Gidrotek. Stroitel.*, no. 2, 1940.

156. Voshchinin, A. P. (Вощинин, А. П.): Flow of Ground Water in the Body and Base of an Homogeneous Earth Dam with an Horizontal Underdrain with a Finite Depth of Permeable Material, *DAN*, vol. 25, no. 9, 1939.

157. Voshchinin, A. P.: Computations for an Homogeneous Earth Dam Constructed on a Permeable Base, *PMM*, vol. 12, 1948.

158. Vreedenburgh, C. G. F., and O. Stevens: Electric Investigation of Underground Water Flow Nets, *Proc. of Intern. Conf. Soil Mech. and Foundation Eng.*, vol. 1, Cambridge, Mass., 1936.

159. Vreedenburgh, C. G. F.: On the Steady Flow of Water Percolating through Soils with Homogeneous-Anisotropic Permeability, *Proc. of Intern. Conf. Soil Mech. and Foundation Eng.*, vol. 1, Cambridge, Mass., 1936.

160. Weaver, W.: Uplift Pressure on Dams, *J. Math. and Phys.*, June, 1932.

161. Wyckoff, R. D., and D. W. Reed: Electrical Conduction Models for the Solution of Water Seepage Problems, *J. Appl. Phys.*, vol. 6, 1935.

162. Wylie, C. R., Jr.: "Advanced Engineering Mathematics," McGraw-Hill Book Company, Inc., New York, 1960.

163. Zamarin, E. A., and V. V. Fandeev (Замарин, Е. А. и В. В. Фандеев): "Hydraulic Structures," Sel'khozgiz, Moscow, 1960.

164. Zhukovsky, N. E. (Жуковский, Н. Е.): "Collected Works," Gostekhizdat, Moscow, 1949–1950.

Index

DE RE METALLICA, Georgius Agricola. The famous Hoover translation of greatest treatise on technological chemistry, engineering, geology, mining of early modern times (1556). All 289 original woodcuts. 638pp. 6¾ × 11.
60006-8 Clothbd. $15.95

SOME THEORY OF SAMPLING, William Edwards Deming. Analysis of the problems, theory and design of sampling techniques for social scientists, industrial managers and others who find statistics increasingly important in their work. 61 tables. 90 figures. xvii + 602pp. 5⅜ × 8½.
64684-X Pa. $14.95

THE VARIOUS AND INGENIOUS MACHINES OF AGOSTINO RAMELLI: A Classic Sixteenth-Century Illustrated Treatise on Technology, Agostino Ramelli. One of the most widely known and copied works on machinery in the 16th century. 194 detailed plates of water pumps, grain mills, cranes, more. 608pp. 9 × 12.
25497-6 Clothbd. $34.95

LINEAR PROGRAMMING AND ECONOMIC ANALYSIS, Robert Dorfman, Paul A. Samuelson and Robert M. Solow. First comprehensive treatment of linear programming in standard economic analysis. Game theory, modern welfare economics, Leontief input-output, more. 525pp. 5⅜ × 8½.
65491-5 Pa. $12.95

ELEMENTARY DECISION THEORY, Herman Chernoff and Lincoln E. Moses. Clear introduction to statistics and statistical theory covers data processing, probability and random variables, testing hypotheses, much more. Exercises. 364pp. 5⅜ × 8½.
65218-1 Pa. $8.95

THE COMPLEAT STRATEGYST: Being a Primer on the Theory of Games of Strategy, J.D. Williams. Highly entertaining classic describes, with many illustrated examples, how to select best strategies in conflict situations. Prefaces. Appendices. 268pp. 5⅜ × 8½.
25101-2 Pa. $5.95

MATHEMATICAL METHODS OF OPERATIONS RESEARCH, Thomas L. Saaty. Classic graduate-level text covers historical background, classical methods of forming models, optimization, game theory, probability, queueing theory, much more. Exercises. Bibliography. 448pp. 5⅜ × 8¼.
65703-5 Pa. $12.95

CONSTRUCTIONS AND COMBINATORIAL PROBLEMS IN DESIGN OF EXPERIMENTS, Damaraju Raghavarao. In-depth reference work examines orthogonal Latin squares, incomplete block designs, tactical configuration, partial geometry, much more. Abundant explanations, examples. 416pp. 5⅜ × 8¼.
65685-3 Pa. $10.95

THE ABSOLUTE DIFFERENTIAL CALCULUS (CALCULUS OF TENSORS), Tullio Levi-Civita. Great 20th-century mathematician's classic work on material necessary for mathematical grasp of theory of relativity. 452pp. 5⅜ × 8½.
63401-9 Pa. $9.95

VECTOR AND TENSOR ANALYSIS WITH APPLICATIONS, A.I. Borisenko and I.E. Tarapov. Concise introduction. Worked-out problems, solutions, exercises. 257pp. 5⅜ × 8¼.
63833-2 Pa. $6.95

CHALLENGING MATHEMATICAL PROBLEMS WITH ELEMENTARY SOLUTIONS, A.M. Yaglom and I.M. Yaglom. Over 170 challenging problems on probability theory, combinatorial analysis, points and lines, topology, convex polygons, many other topics. Solutions. Total of 445pp. 5⅜ × 8½. Two-vol. set.
Vol. I 65536-9 Pa. $5.95
Vol. II 65537-7 Pa. $5.95

FIFTY CHALLENGING PROBLEMS IN PROBABILITY WITH SOLUTIONS, Frederick Mosteller. Remarkable puzzlers, graded in difficulty, illustrate elementary and advanced aspects of probability. Detailed solutions. 88pp. 5⅜ × 8½.
65355-2 Pa. $3.95

EXPERIMENTS IN TOPOLOGY, Stephen Barr. Classic, lively explanation of one of the byways of mathematics. Klein bottles, Moebius strips, projective planes, map coloring, problem of the Koenigsberg bridges, much more, described with clarity and wit. 43 figures. 210pp. 5⅜ × 8½. 25933-1 Pa. $4.95

RELATIVITY IN ILLUSTRATIONS, Jacob T. Schwartz. Clear non-technical treatment makes relativity more accessible than ever before. Over 60 drawings illustrate concepts more clearly than text alone. Only high school geometry needed. Bibliography. 128pp. 6⅛ × 9¼. 25965-X Pa. $5.95

AN INTRODUCTION TO ORDINARY DIFFERENTIAL EQUATIONS, Earl A. Coddington. A thorough and systematic first course in elementary differential equations for undergraduates in mathematics and science, with many exercises and problems (with answers). Index. 304pp. 5⅜ × 8¼. 65942-9 Pa. $7.95

FOURIER SERIES AND ORTHOGONAL FUNCTIONS, Harry F. Davis. An incisive text combining theory and practical example to introduce Fourier series, orthogonal functions and applications of the Fourier method to boundary-value problems. 570 exercises. Answers and notes. 416pp. 5⅜ × 8½. 65973-9 Pa. $8.95

THE THOERY OF BRANCHING PROCESSES, Theodore E. Harris. First systematic, comprehensive treatment of branching (i.e. multiplicative) processes and their applications. Galton-Watson model, Markov branching processes, electron-photon cascade, many other topics. Rigorous proofs. Bibliography. 240pp. 5⅜ × 8½. 65952-6 Pa. $6.95

AN INTRODUCTION TO ALGEBRAIC STRUCTURES, Joseph Landin. Superb self-contained text covers "abstract algebra": sets and numbers, theory of groups, theory of rings, much more. Numerous well-chosen examples, exercises. 247pp. 5⅜ × 8½. 65940-2 Pa. $6.95

GAMES AND DECISIONS: Introduction and Critical Survey, R. Duncan Luce and Howard Raiffa. Superb non-technical introduction to game theory, primarily applied to social sciences. Utility theory, zero-sum games, n-person games, decision-making, much more. Bibliography. 509pp. 5⅜ × 8½. 65943-7 Pa. $10.95